SEO 3.0

The New Rules
of Search Engine Optimization

Michael J. Meyer

SEO 3.0 – The New Rules of Search Engine Optimization

ISBN: 978-0-9834007-0-7

Credits

Written by Michael J. Meyer

Edited by Amanda Bradbury & Robert Ackert

Cover Design by Christal Vass

Creative Consulting by Travis Boatright

Disclaimer

Acknowledgments

Just as in managing a successful company, writing a book worth reading is a team effort. That said; I'd like to thank everyone in that team, from colleagues to clients. The list is too long to do each contributor justice so I'll keep it to the basics…

I'd like to thank everyone at NetSearch Direct for their willingness to provide endless proof-reading, graphics skills and – above all else – brutal honesty when I needed it. And special thanks to Mike Small, my SEO Manager, for his unfaltering care of clients and ability to keep one hundred plus websites ranked at the top of Google, Yahoo! and Bing for their valuable keywords.

More than one hundred clients also made this book possible by trusting my company to bring online success to theirs. Without their initial leap of faith, none of us would have experienced the great success we have, and the information in this book would be interesting theory rather than verified fact.

I'd also like to thank Biff Wittkamp and Dave Jones, my two business partners of almost twenty years. They have supported me in the development of NetSearch Direct, while guiding our other direct mail businesses to success. To my colleagues at Virginia Council of CEO's, each running a successful company yet still finding the time to give me valuable input and much needed feedback.

And last, but far from least, I need to thank my wife Helene and daughters Kendall and Logan whose support and understanding, of my time sequestered to the den, made this book a reality.

Feedback

While writing this book, I sent unedited, "work in progress" copies to a handful of respected leaders from various industries whose opinions I knew would help make this a better book.

Here is their feedback...

"NetSearch Direct helped us understand the value of SEO and helped us move from not being optimized at all to be ranked on search engines. Now all aspects of our marketing efforts incorporate SEO strategy."

Heather Gough
Director of Marketing & Communications
FranNet

****

"Whether you are managing your website in house or through an outside vendor - SEO 3.0 provides extremely valuable guidance for C level executives and small business owners alike."

Joe Bourdow, CFE
Senior Advisor
Valpak Direct Marketing Systems, Inc

****

"I've known Mike for seems like a hundred years... who knew he had a book inside him? No techno-babble here. Finally a book that explains SEO in language that a businessman can understand. Great job Mike!"

George Boatright
Chief Information Officer
Media General

****

"These guys get it and now you can too. SEO 3.0 is a must read for any business owner who wants a practical understanding of how search engine optimization can impact their business."

Bryan Jones
President
CWDkids

****

"As President of one of the oldest ongoing SEO companies in the United States I can say without question that SEO 3.0 will forever change the way you look at your search engine optimization efforts.

Written by an executive SEO expert for other executives, SEO 3.0 is aptly named as it catapults the reader over out-of-date concepts, blazes a trail through bleeding edge innovations and breaks down the techno-barriers that keep most managers guessing about what is really going on behind the curtain.

Now anyone savvy enough to invest two hours reading this concise missing manual will know without question what new rules have to be followed to earn the top spots in this new era of search.

If you are in any way responsible for the online success of your company, you cannot afford to pass this by."

R. C. Burgess
President
DotComPirates.com

****

"Reading SEO 3.0 will change the way you view search engine optimization and put you on the way to making your website a "lean, mean lead producing machine". I know because Mike and his team have implemented many of the principles included in this book and have produced outstanding results for my company in a very competitive industry."

Dusty Rhodes
Vice President
Marketing and Business Development
SMARTBOX

Foreword

SEO. Often hyped, certainly oversold, yet incredibly valuable.

Is SEO Internet black magic? Or is it something that is predictable? The answer to those questions depends mostly on your personal experience and planning. If you don't arm yourself with knowledge, and opt to go with a discount provider, then certainly the 'black magic' claims will sound appealing.

In 1999 some people claimed that the SEO field was doomed to fail. During 2003 I got my start in the SEO space & Google did an update named 'Florida.' It drastically altered the result set, and once again the claims of 'SEO is dead' resurfaced. In the years since there has been Google Universal results, search personalization, localization, Google Instant, and a variety of other sea shifts in the space. Each came with the claim that 'SEO is dead.'

Google makes about 500 changes to their algorithm every year. Yahoo! and Ask folded their search engines in 2010. Bing came from nowhere to number 2 in a couple years, as competitors gave up on the space. New search engines like Blekko offer another voice in search. The players in the search game change rapidly, as does the structure of the web.

SEO grows more complex every year. A lot of the basic concepts stay the same, but where the algorithms change some people fall behind and others get pushed ahead. Arming yourself with knowledge is the key to ensuring you are in the group who keeps growing at 20% a year in spite of whatever seems to be happening in the 'real' economy.

Like a cat with 9 lives, SEO keeps 'dying'. People write it off & miss the opportunity, which is a good deal for everyone who takes the time to study the field and leverage it to grow their business. One of the best ways to learn is to get your hands dirty setting up some test sites and tracking how they evolve over time. But many people give up before they have a chance to succeed as the barrier to entry keeps growing. Often people are literally weeks to months from turning the corner, but quit. They didn't know how close they were to succeeding and lacked the confidence to keep investing in something which is 'dead.'

When people search they are telling you exactly what they want. It's not a model built on bombarding millions of people and hoping a few thousand care. The efficiency is staggering - just look at Google's revenues. Many businesses which get only a few dozen to a few hundred targeted search visits a day are thriving. Others have come from quiet and humble beginnings in a basement to building businesses that are larger and more profitable than 'real' (businesses which are claimed to be superior.)

By reading a leading book about SEO you not only give yourself the baseline knowledge of where search is today, but you also give yourself key insights into where it is headed. With that comes the confidence to invest aggressively in the most powerful marketing channel created in the past 100 years.

— *Aaron Wall; founder of SEObook.com, and respected SEO Expert*

Author's Note

We all have our favorite movies...right? There could be many reasons why a particular movie makes our list of personal favorites. Usually there is something that reaches out, grabs us and makes us think. It could be the storyline, a performance by an actor or a particular quote. Such is the case for me with _The Patriot_, a 2000 film depicting the story of a farmer swept into the American Revolutionary War when his family is threatened. The reluctant hero is played by Mel Gibson. Heath Ledger (who plays his son) and Tom Wilkinson (British General Lord Cornwallis) also star in the film. I love this film for many reasons, but a dialogue between Lord Cornwallis and his aide has stuck with me for 10+ years.

One of the last scenes of the movie features the Siege of Yorktown - the last major battle of the American Revolutionary War. The upset of all upsets is about to take place. The British, at the time considered to be the most powerful nation in the world, were about to surrender to an army of blacksmiths and farmers; the Americans. In this scene, as Cornwallis' aide is beseeching him to surrender, the bewildered British General laments... _"How could it come to this? An army of rabble. Peasants. Everything will change. Everything has changed."_

Change. That's what this book is all about. As a veteran of 29+ years in the advertising industry I can say that never have I witnessed such a period of change in this profession, as we have experienced in the last five years. A convergence of factors has led to this – creating an almost perfect storm that has revolutionized the advertising and marketing industry; much like the "perfect storm" of events that prompted a group of ill-equipped "Colonials" to stand up to the most powerful force of the day and change the world forever.

Can you say *dotcom*? This environment of change all started with the Internet (more specifically the web) and subsequent creation of what is now known as Search Engine Marketing. Believe it or not there was a period of time where the Internet existed without Google. However, Larry Page and Sergey Brin - the "Benjamin Franklin" and "Thomas Jefferson" of the impending marketing revolution – certainly changed the game forever with the launch of Google in 1998. Yet, if you were like me you were probably too busy trying to take advantage of one of the most prolific economic periods of the 20th century (March 1991 to March 2001) to understand the significance of this event. It really wasn't until after Google went public in August 2004 that many business owners took notice that the game was changing. Even then, most of us could not have imagined what was ahead.

In late 2004 and early 2005 the search engines (primarily Google, Yahoo! and MSN) refined their geo-targeting capabilities. Now *local companies* could much more easily utilize pay-per-click to direct potential customers to their business. *Local Search* (where a specific geography was included as part of the search phrase) really took off. It had been this way since the late 1990's for Search Engine Optimization, but now even more so as an influx

of post dot-bomb SEO professionals also began to fully capital-
ize on the ability to optimize a website for geographically specific
searches. A whole new world was opening up to consumers
seeking an easier and quicker way to find information on *local*
products and services.

Local Search contributed to an exponential increase in the
consumer's utilization of the search engines. Now the search
engines offered a highly targeted and measurable – not to mention affordable – way to reach a particular market. While this all
sounds great (and in hindsight it was), the advent of this new
marketing alternative was at the same time confusing to many
business owners and conventional marketing companies. Something was wrong. Traditional marketing was becoming less and
less effective and this new "search engine" marketing was more
and more confusing.

For me that is when Cornwallis' words rang true. *"Everything will
change. Everything has changed."* The consumer was now in charge
and that was a big change indeed.

Whether businesses realized it or not the consumer was now
choosing how and when they were going to be entertained and
what messages were delivered to them. The Internet offered
consumers an unprecedented opportunity for this. Media had
changed dramatically in the twenty plus years since I started my
marketing career. For example, in the early eighties there were
relatively few choices regarding electronic media. Cable TV was
in its infancy. Essentially, three to four television networks and a

handful of local radio stations were the electronic media choices available to consumers and advertisers. Newspapers were thriving and the yellow pages had a monopoly on "search". Outdoor and direct mail basically rounded out a business owner's menu of marketing alternatives. And with the exception of direct mail (in particular direct mail coupons), these traditional advertising vehicles offered little measurability and even less accountability.

Who was in charge? Up until the dawn of the twenty first century it was most certainly the media companies. Millions were made by media companies as long as the alternatives for news, information and entertainment remained somewhat limited for consumers and businesses. However, as the Internet and the search engines matured these "alternatives" dramatically increased. And the First Commandment of Advertising *"He Who Delivers the Audience Will Be Delivered the Cash"* kicked in. In February 2006 Jupiter Research reported that online users spent as much time surfing the web as they did watching television, and more time online than they spent on all other media combined. Six months prior to that in August 2005 Jupiter Research had predicted *"Internet Ad Spend to Reach $18.9 Billion By 2010"*. They were close – only off by 33%, as ad spending on the Internet reached approximately *$25 billion in 2010*. The worm had definitely turned. But wait, there's more…

The winds of change were about to provide us the final ingredient needed to create that "perfect storm" which was to forever change the way businesses go to market. Much like it takes just the right set of factors to combine to churn a tropical storm into a hurricane, such was the state of the advertising industry in 2008. Let's review.

Things were going swimmingly for traditional media companies until the Internet started to pick up steam after the turn of the century. Consumers began to utilize online search at an exponential rate as Local Search gave them the ability to easily and quickly find the answers they needed. As Americans spent less time consuming traditional media and more time online, not surprisingly, the money (i.e.ad spending) followed.

The final piece that stirred this media storm into a full out revolution was simply the Economic Downturn of 2008. The "worst recession since the depression" rocked the world of U.S. businesses and has forever changed their view of advertising. Branding, Cost Per Point and Nielsen Ratings are out. Return-On-Investment, Pay-Per-Click and Measurable Results are in. The Economic Downturn/Recession of 2008-2010 required business owners to closely scrutinize every expense. Conveniently, the Internet gave them the tools to make media companies accountable.

Winston Churchill said, "There is nothing wrong with change, if it's in the right direction." I submit that this aptly describes the opportunity presented by the state of the advertising industry today. Search Engine Marketing - and in particular Search Engine Optimization – offer the most accountability and best ROI among all the choices on a marketer's menu. So in this case, for business owners, change is good. Those who understand, embrace and implement the new rules of this advertising game, not to mention the *New Rules of Search Engine Optimization*, will win. And that, my friend, is what this book is all about.

Table of Contents

Feedback 5

Author's Note.. 11

Introduction.. 19

What to Expect 20

Getting the Most Out of This Book.. 23

SEO by Any Other Name.. 25

Part One: SEO 1.0 and 2.0.. 29

SEO 1.0 (1995 to 1998) 31

SEO 2.0 (1999 to 2010) 33

SEO 3.0 (End of 2010 and on…) 35

SEO 1.0: The Methods Behind the Madness 36

SEO 2.0: The Wild West of the Internet 43

Classic Rules of SEO.. 53

An Added Edge 58

SEO 3.0: The New Rules 59

The Changing Face of SEO 61

How Are You Evolving? 63

Forget Clicks - Get Leads.. 65

The Big Three Are Now the Big Two...................................... 67

Can Google Read Your Mind? 71

Getting Local With Google Places... 77

Google Place Search.. 83

Google Instant Preview 85

Keeping Track with Google Analytics 89

Goals and Conversions 93

Get Cracking on Call Tracking 97

Should You Commit to Submit? 101

Make SEO Copywriting a WOW! 103

Not All Keywords Are Created Equal 105

A Note on Competition 109

Change Happens...Deal With It 111

Get Social and Be Famous 113

Let the Blogging Begin 115

Go Mobile 117

Invent What You Cannot Find 119

Summary 122

Before You Invest a Dime 129

BONUS

Free *Search Engine Snapshot Report* Request Form 191

Introduction

This is not another "how to" SEO book, but rather a "what if" book... *What if* you knew <u>exactly</u> what happened to make your (and your competitors') search engine rankings rise or fall? *What if* you could anticipate (even just a little) the major changes Google, Yahoo!, and Bing were planning weeks before your competitors? Learn the **new rules** and you will be able to do all of this and more.

Written by management for management, *SEO 3.0* is about search engine optimization evolution and how to benefit from *any* change that comes along, at *any* time. It is a chronicle of the most important search engine changes in a decade as well as an expert-led guide to help you prepare for what's next.

In short, this book is about how to win in the new era of search engine optimization (SEO 3.0) that came about after an entire decade's worth of changes were crammed into three short months at the end of 2010.

That said; it is not so much for those turning the nuts and bolts of your online marketing machine, as for those responsible for making sure it runs at its best and does everything it needs to.

Quite simply, if you are in any way responsible for your company's online success, this book is for you.

What to Expect

It's always nice to have a heads-up before you dive into something new. Of course, discovering information that will change your company's "life" is certainly no exception. That said, here's what you can expect from this book…

Expect a quick and easy read. This book is short for a reason. You don't need a ton of detail. All you really need is some solid information and insights on those crucial elements that play the greatest deciding factors in the world of search. By the same token, let's leave the techno-babble to those who are impressed by their own big words and lengthy commentary. We don't need it. It's a waste of time, and your time is valuable.

Expect very recent information to be covered. The whole point of this book is to put you ahead of the curve. Some of the topics covered are still in beta as of January 2011, which puts you way ahead of your competition.

Expect to read some sections more than once. It might be concise but there is still a lot of information to digest. Don't be concerned if you find yourself coming back to check a fact or rethink a decision. The most successful business people in the world are those who follow the simple rule of 1, 2, 3…

1. *Plan*
2. *Review*
3. *Execute*

Expect to disagree with some of what we say. It's normal to question new information, especially if you have already paid someone else who has told you something different. Just keep an open mind and consider each point on its own merit. We've made a lot of people a great deal of money by using the very same information you currently hold in your hands.

Expect to get answers. The worst thing about failing is simply not knowing why you failed. Slightly less worse (I just can't bring myself to say 'better' about this situation) is being misinformed, accidentally or intentionally, about *why* you failed. Either way, you can expect to keep repeating mistakes until you know what went wrong the first time around.

Consider this… More than five billion (yes, 5,000,000,000) web pages' rankings changed literally overnight in late 2010, and very few people know why – even today. By the end of this book, you will know exactly what happened and how to anticipate and even prevent similar occurrences in the future.

Expect help whenever you need it. We know that your time is valuable and that sometimes you just have a question or two about strategy. Call us anytime. We are happy to hear from you and even offer hourly consulting services for those more in-depth issues holding you back from greater success.

You can reach us at 804-228-4400 Mon – Fri 8:30 a.m. to 5:30 p.m. Eastern Time. Or feel free to shoot me (Mike Meyer) an email at MikeM@NetSearchDirect.com anytime.

By the way, these are not just words intended to be helpful. I really do want to hear from you. As comprehensive as I have tried to make this information, some of it can still be confusing, and even overwhelming at times. I am here to help.

Getting the Most Out of This Book

I cannot stress enough how important the information contained within these pages can be to your future online success. Of course, it takes more than just reading – or even understanding. It takes **planning** and **action** to succeed. You already know this of course, but in this world of instant gratification sometimes a gentle reminder can be a good thing.

It goes without saying that you might not opt to use every strategy in this book. Just know that each one you *do* use will help your overall campaign in multiple ways. And as a nice bonus, incorporating all aspects creates a *whole that is greater than the sum of its parts* as they complement and build on each other. As you progress into the action stages, you'll see what I mean.

Regarding the layout of the book goes, it's pretty straightforward. There are just a few major sections as follows…

Part One: SEO 1.0 and 2.0

The initial portion of this book provides a quick summary and timeline of SEO 1.0 and 2.0. It's a basic primer explaining what shaped the way search engines, and SEO, work today.

But it is even more. You will also see how modern "black-hat" tactics evolved, step-by-step, and how to identify any of these techniques that could get your site permanently banned from Google, Yahoo! and Bing.

Part Two: SEO 3.0

The era we call SEO 3.0, began in late 2010 – and turned the world of search engine optimization upside down.

This area of the book deals with the most recent and world-altering aspects of search engine modifications. You will see exactly what changed and how to plan an extremely profitable SEO campaign by following a few simple rules.

It would take a much bigger book to address every aspect of a campaign, and there are already dozens of books on the shelves that do that now. Our goal is to provide the missing information that will help you trample your competition TODAY by planning for those inevitabilities they do not.

Part Three: SEO FAQ

As a business owner or manager whose success depends on the consequences of your search engine optimization results, you undoubtedly have questions. This detailed Frequently Asked Questions (FAQ) section has plenty of answers.

But not just any answers. These FAQ's are not simply random throw-ins. Each question within these pages was asked by a busy professional whose success depends on their website's ability to be found easily on all major search engines.

That's about it. Enjoy the book and feel free to call us at 804-228-4400 with any questions.

SEO by Any Other Name

Search engine optimization, or SEO, goes by many names. Some of the more popular are *website optimization, search engine placement, search engine marketing, organic optimization, and natural optimization.* Of course, those outside of the US, typically spell it "optimisation", with an "s", which further adds to the mix.

Then again, many people use the term SEO to describe a search engine *optimizer*, or professional.

For simplicity sake let's stick with "SEO" as meaning search engine optimization and "SEO'er" as referring to one who performs search engine optimization tasks.

Okay, that tells us **what**, but **why**? I mean, why bother with SEO in the first place?

Well, I find profit is probably the best motivator with reputation protection as a close second. Allow me to elaborate…

Profit: Nearly every credit card carrying consumer in the United States has access to the Internet - at home, work or through their Smartphone. Even if they don't buy your product or service online, they can still learn about it from the web.

But this is old news. If you picked up this book, you probably already know this. Here's a tidbit that might be new.

Reputation: Whether you sell on the Internet or not, if you have customers, they are probably talking about you.

What I mean to say is that someone else is telling your story; good or bad. If it's good - congratulations. If it's bad then you need your story, as you tell it, to appear above all others when someone searches Google.

This addresses two of the most common questions on any executive's mind when considering SEO. The third of course, relates to cost and value.

You are the only one who can place a value on being found before your competitors when a potential customer is searching. I can however, offer one statistic and one consideration.

Statistic: As of 2009, approximately ***70% of all Internet users completely skip over the paid placements on search engines and go straight to the organic listings (the SEO stuff).*** This was up from the mid 60's just a couple years before. In other words, they trust paid ads less (and less) than regular websites. Not really a big surprise when you put it into context.

Equate this to reading a magazine. What would you trust more, an article (even a press release you wrote) singing the praises of Product X or an ad Product Y paid a fortune to put in front of you? Personally, I would go with Product X every single time.

But let's not leave it to the imagination...

Paid Ad (pay-per-click landing page, etc.):

Save Money Now!

Buy a "Comfort Select Super Sleeper" mattress from a Comfort Select factory location to save time and money.

Here are just a few of its benefits:

- "Never flip" pillow-top
- Stuffed with 100% allergy-free foam
- Proudly made in the USA

Article Excerpt (press release, web page content, etc.):

We spend about one-third of our lives sleeping but so few people take the time to select the ideal mattress it's actually quite sad. They work, day in and day out, just to come home and deny their body the rest it needs to help keep them strong, healthy and active.

Knowing that it's usually a combination of cost, time, and convenience that stops people from getting the rest they deserve, **Comfort Select** has over 250 factory-direct outlets in the US, each with a sample of every mattress type available, including the newest Comfort Select Super Sleeper.

It has all the comforts that you'd expect and more, such as:

- "Never flip" pillow-top
- 100% allergy-free foam
- Handcrafted in the USA by experienced sleep experts

Consideration: What is the value of a sale? It does not matter whether you are selling a product online or offline.

How valuable is it to place your business in front of a prospect who has taken the time to go online, search for a particular product or service and then click on your website to find out more about how you can be of help to them?

To put it another way, what if you *only* paid for television, radio or newspaper ads which were viewed by consumers who had indicated they were interested in what you had to sell? Wouldn't that provide you with the optimum ROI? Well that is precisely the opportunity SEO offers to businesses today.

Part One: SEO 1.0 and 2.0

In this book, I refer to SEO in stages for the purpose of clarity. SEO 1.0, 2.0 and 3.0 are not *official* terms you are likely to find universally across dozens of books and hundreds of websites. I am merely borrowing the concept of "web 2.0" which, as I think of it, was created for the same basic reason. That said; let's jump right in and find out how it all began, how it continues to evolve, and, of course, how you can profit from it.

SEO 1.0 is where it all began.

SEO 2.0 is what most of us know today.

SEO 3.0 is what we need to know to succeed tomorrow.

While the bulk of material covered here is about what we need to know for future success, we want to spend just a little time discussing the early days of SEO.

Believe it or not, many so-called professional search engine optimization "experts" are still using many of these (now banned) techniques and charging an arm and a leg for work that could get your site permanently barred from major search engines like Google. Yes, seriously.

Knowing this, fifteen minutes spent learning the old-school stuff is a good investment. We will start with a chronological outline of each of these stages. Then we will get to the good stuff – real world examples or characteristics of SEO 1.0, 2.0 and 3.0. So let's do it…

SEO 1.0 (1995 to 1998)

To put things into perspective let's look at the very beginnings of SEO. For the most part, search engine optimization started back in 1995 when the web was first really being used as a way to promote business and help make sales. Back then, most websites were not much more than online brochures and animated GIF's were considered cutting edge technology.

Google was still years away from development and the biggest search engine was not a search engine at all. It was a directory called, "Yahoo!", invented by two college students (Jerry Yang and David Filo in January 1994) who basically just compiled a great big bunch of links to their favorite websites and made the list searchable.

When Yahoo! was incorporated and released to the masses in 1995, search engine optimization (more commonly known as SEO) was not far behind. Of course, back then it was called "website optimization," but that's another story.

From 1995 to 1997, Yahoo! was the unchallenged champion of the business search world while AOL Search was one of the biggest players in the consumer market. Why? Because seventy percent of the home user market was on AOL's dial-up Internet service and the default homepage was AOL.com with its (then ground-breaking) simple search box.

Speaking of ground-breaking… In 1996, while attending Stanford University as Ph. D candidates, Larry Page and Sergey Brin collaborated on a search engine called BackRub. The company started picking up traction in 1997 and Page/Brin decided that their search engine needed a new name; Google. Things really took off in August 1998 after Andy Bechtolsheim (co-founder of Sun Microsystems) dropped $100,000 in venture funds into this project. This was about the time Microsoft threw its hat into the ring with MSN Live Search.

One of the most important things to remember about this timeframe was that it was the very beginning of search engine competition and bigger was always considered better. Search engines like Google and directories like Yahoo! were judged more by the number of web pages they had in their catalog than by the quality of search. This was back when there were hundreds of thousands and, by 1998, millions of pages on the web. Any page was a novelty and coveted by all search engines.

SEO during these days was a pretty simple affair, but we will get into that soon enough.

SEO 2.0 (1999 to 2010)

If you felt the need to pin a start-date for the era of SEO 2.0, I would have to say it began in 1999 when search engines knew their existence was not just a fad - and that they could consistently make a lot of money from banner ads, paid placement as well as a dozen other ways. Now that search had a concrete value, every search engine and directory wanted the biggest possible piece of the pie.

In the early days of the SEO 2.0 era, from about 1999 to 2002, there were traditionally a group of "top ten" search engines. These always included Yahoo!, Google, and MSN (now Bing) while the other seven came and went. These included AltaVista, Alltheweb, Dogpile, AOL Search, Earthlink Search, Inktomi, Looksmart, Search.com, Ask Jeeves (later Ask.com) and others long forgotten.

This is also when the "Internet bubble" burst and people were discovering that paying one million dollars to air a 30 second home-made, camcorder created dotcom Super Bowl commercial was not the best use of investors' money - and that even Internet startups needed to be held accountable for that thing called profit. Dark times indeed for the dreamers and schemers who thought the web was their golden parachute.

Soon after the post bubble shakeout, around 2002, reality set in about the real potential of the Internet and search engines got even more competitive. It was no longer about getting as many pages listed as possible. It was now about being a curator rather than a warehouse, and presenting only the very best search results to those profit-generating visitors who now had multiple options when it came to search.

By 2005, major search engines did all they could to limit the pages included in their index and even started charging for "guaranteed" or "expedited" inclusion. But that's not all. They also banned the common methods of SEO from just a few years earlier and started investing fortunes in new technology to improve search results and catch anyone using the recently barred optimization tactics. But you will learn all about that in a few minutes.

The SEO 2.0 era continued through to the latter half of 2010. Then in August through December, about a decade's worth of change happened and changed the world of search in the biggest ways yet...

SEO 3.0 (End of 2010 and on...)

In August 2010 Bing and Yahoo! finally put into place what they had been threatening for more than a year – combined search results.

Now, Yahoo! displays Bing's results while retaining its own brand. This in itself is a big deal – especially for anyone who spent money optimizing for Yahoo! over Bing in recent months – but even more so because it was the first major change that spurred the SEO 3.0 era.

Before most business owners had a chance to digest the news about Yahoo! and Bing, Google started a series of upgrades that changed nearly every aspect of the existing search game as we knew it. They barraged more than five billion web pages with altered rankings by implementing five major changes in just a few short weeks. That's what the bulk of this book focuses on, but it's worth your time to read the next two short sections on the methods behind SEO 1.0 and 2.0. This is definitely one of those areas where what you don't know can, and will, hurt you.

SEO 1.0: The Methods Behind the Madness

The early days of search engine optimization can be likened to a giant lab filled with hundreds of mad scientists who could experiment with anything and get away with it. They had no concerns over breaking rules because there simply were no rules to break.

Scary Fact: One enterprising self-proclaimed "Internet guru" sold a 3-ring binder full of these little tricks – for $300 a pop! And he could not keep up with demand, selling thousands upon thousands and earning millions.

Scarier Fact: Even now, in 2011, one out of every five new client sites we evaluate contain one or more of the following black-hat optimization techniques – all of which were banned a decade ago or earlier.

Here is some of what these evil SEO geniuses came up with:

Keyword Jacking

Back in the 1990's, search engines gave a lot of weight to anything in a Meta tag (website HTML code that human visitors don't see). Back then there were more "get rich quick" websites ranking for the keyword "Pamela Anderson Lee" than there were *"Baywatch"* fan sites. Webmasters and SEO people just did not care. Any hit was a good hit – no matter what pretence the visitor arrived upon. Try that now and you can actually get sued.

1.0	Don't Borrow Keywords	Classic Rule

Whether they are in your main page text, Meta tags, Titles, etc - Do NOT "borrow" any keywords that you have no legal right to use. This is especially true of competitor's names or brand names that are subject to intellectual property law. This is a BIG deal. You can be sued for using a competitor's name or infringing on their intellectual property.

Keyword Spamming

After the success of stuffing irrelevant keywords into Meta tags, webmasters decided they could really get a boost from stuffing keywords into the on-page text. And, they were right! This became known as keyword spamming. It was common to see pages with two or three paragraphs of real text followed by the same keyword repeated 100, 200, even 500 times.

1.0	Stay Keyword Natural	Classic Rule

"Keyword natural" is a quick way of saying you should use all keywords genuinely, as they would normally appear in conversational writing.

Using any keyword a few times per page, only where they make sense, is the way to go. There is no need to cram more in than are really needed. In this case, less is more.

Invisible Text

This was extra sketchy from the very beginning. Here's how it worked... The webmaster/SEO'er would add a bunch of keyword laden text to the page and make it the same color as the background, thus seemingly invisible to the visitor. White colored text on a white background was the most common.

This was one of the original "black hat" techniques, and it evolved every time the search engines caught up – for years! This was the turning point when search engines decided they had had enough of unscrupulous optimization tactics. Here's what happened...

As you already know, the search engines said "enough of this" and banned the use of so called "invisible text". Google and the other big players initially thought this would be enough to reduce the majority of the abuse. Boy, were they in for a surprise!

Greedy webmasters and SEO'ers decided to play it fast and loose by looking for loopholes in the color palette. And they found plenty. They knew that search engines cannot 'see' color. They can only read color codes. And with 256,000 colors and shades in the (then) current spectrum, many decided to get around the system by using a color that was simply a shade off, but virtually indistinguishable to the human eye. A common example seen in millions of pages was two shades of green, identified by hex color code ID's #66FFFF and #33FFFF. They look 99.9% identical to the human eye but search engines interpreted them as totally different colors – for a while.

Once major search engines caught on to this scam, they started spending hundreds of thousands of dollars in developing technologies that could tell when invisible text was being used based on spectrum analysis.

1.0	Visibility = Transparency	Classic Rule
	Search engines can tell when you are trying to keep things hidden from your visitors, so don't bother. Be transparent. Be REAL. They see "invisible" text as cheating the system and will punish your site if they catch you at it. And don't think for a minute that your visitors are oblivious. This trick is so old that just about everyone knows how to spot it by highlighting big blank areas and watching that hidden text come to life.	

Tiny Text

Not to be outdone when search engines started battling invisible text methods, some SEO people said "why not make text too tiny to identify?" And that's just what they did. After all, if you make text in a 'one point' size and an even remotely similar color to the background, you can add hundreds of keywords and nobody will ever know.

Then some even more enterprising webmasters decided to let the tiny text stand out with a bold color and put it into easily formatted tables so the text itself looked like decorative squiggly borders. This worked for years. After all, why would a search engine bother looking at point size? A character looked the same to a 'bot' whether it was 1 pt or 12 pt.

1.0	Say it BIG. Say it Loud!	Classic Rule

Don't ever use tiny text on a website. It looks awful when people think you are hiding something. And once they figure it out, that is their ONLY area of focus.

If you have something to say, say it big and loud. Even your copyright and disclaimer should be easy to read for anyone who wishes. In fact; this can be a great place for that final keyword.

Moving On...

The real take-away here is that search engines spend a fortune trying to upset the efforts of black-hatters (unethical SEO'ers) and once caught, they have no problem banning it for LIFE.

Okay, but why? Larry Page, co-founder of Google pretty much summed it up when he said "The perfect search engine would understand exactly what you mean and give back exactly what you want." In other words Google believes its success lies in its ability to deliver the absolute best answer to a search query faster and better than anyone else. Anything that gets in the way of this jeopardizes their success.

1.0	Think "Would I Mind?"	Classic Rule

Before doing anything that is remotely questionable - take a moment to actually ask the question "If I were in the search engine's place - investing millions to provide the most relevant search results possible - would I mind if someone did that?"'

Remember: SEO is about making your site the <u>most relevant</u> it can be <u>for targeted keywords</u>, NOT fooling search engines.

Want More?

That covers the major points of SEO 1.0 in a nutshell. There is a good deal more of course, but we covered the most commonly found, and the most dangerous.

Just in case you are curious however, here are some more things from this era, to be on the lookout for...

Auto-generated Content Pages: *This became popular with a product called WebPosition Gold in the late 1990's. Although there were others before that, WebPosition made it possible to create hundreds of keyword-laden pages per day by simply filling in a few fields of data and pushing a button. This tool even allowed the user to select the search engine it most wanted to target for SEO purposes, for each new page created.*

Invisible Links: *We discussed invisible text. This is nearly the same but refers to trying to get added keyword "juice" by placing an invisible keyword in a link so nobody knows it's there.*

Why bother? Because invisible links allow you to promote anything you want without the visitors ever knowing that is what your site is really about.

Misleading Headings: *The H1 through H6 "heading" tags are some of the more powerful on-page optimization attributes so some unethical SEO-ers place deceptive content in these tags in order to attract visitors, links and organic SEO from keyword use.*

Google Bowling: *This is when you intentionally "bowl down" the rankings of websites you don't like by linking to their competitors, whether you like them or not.*

Please Note: Google Bowling is not done for the purpose of helping those sites you link to, but rather hurting the site they will now move down in the organic rankings.

Disposable Domains: *Since most major search engines give priority to exact match domains (those that match the keyword exactly), some SEO'ers will purchase as many of these as possible and build micro-sites with no intention of keeping them up. They simply abandon them once they squeeze every click they can from them.*

This is especially true regarding the latest money-making and/or weight-loss trends. Once the trend, and keywords associated with it, have died down in popularity - the site is abandoned.

SEO 2.0: The Wild West of the Internet

One of the more common descriptions you may have heard about SEO over the past decade is that it was like the "Wild West of the Internet." A bunch of SEO cowboys doing whatever they wanted and hoping they did not get caught by the local authorities. But even if they did, was it really a big deal? Get busted by Yahoo! and you could just double your efforts on Google. Get slapped by Google and you could simply build a new site under a different name. These SEO scam-artists didn't care. The rewards far outweighed the risks – for a while that is.

This is where the plausible deniability of the late '90's was taken over by the obvious "black-hat" illegal/unethical methods (incidentally, "black-hat" is a reference to the bad guys in old westerns).

So what kind of behavior led to the banning of "black-hat" tactics? It can be best described as outright cheating. Here are some samples of what it involved:

The Old Bait and Switch

It's just like it sounds. A webmaster would put up a web page developed completely for search engines (as search engine 'spider bait') and then swap it for an attractive page once it got ranked (the switch).

In those days a bot might not come back and re-spider your page for years, so this was pretty easy to get away with. This, of course, is no longer the case.

2.0	Updating is Not Replacing	Classic Rule

Advanced search engines like Google love fresh content but hate page swapping. ALL major search engines know this is a scam to get long-term high rankings from a temporary page full of "spider bait."

Again, add fresh content and remove stale content, but don't swap out an entire page. Show stability AND improvement.

Link Farms

In about 2003, the importance of linking became more apparent than ever. Google, more so than any other search engine, viewed a web page's value and relevancy to a given keyword by how many links it had. It was basically a popularity contest with each inbound link counting as a vote.

However, back in the early 2000's any link would do the trick. And, reciprocal links (meaning you swap links with another site) were extremely valuable (and genuine). Google and other search engines felt that even a reciprocal linking partnership meant that you had a relationship of value. Then, link farms came along and ruined it for all of the real reciprocal linking relationships.

In a link farm, you simply pay to have a bunch of other link farm members link to your site – and you link to theirs. There is no natural or genuine relationship. For twenty dollars a month you can become link partners with thousands of like-minded sites. And, this became a big no-no on Google's list.

2.0	Link Carefully	Classic Rule

One of the most time-consuming parts of search engine optimization is link building. And for good reason; it's the most powerful aspect of SEO as well.

That said; choose your links well. Never use link farms of any type. They are reciprocal junk that will get you busted and possibly banned for life.

3-Way Linking

When Google caught onto link farms, someone came up with the bright idea to build three-way link relationships. With this method, Google might never know who was involved because you did not link back to the page linking to you. You linked to a third page, which in turn, linked back to the first.

It was a pretty slick arrangement, but since each page in the link circle was in Google's listings, it did not take them long to catch on and just start dinging sites that were obviously using this method.

| 2.0 | Link Naturally | *Classic Rule* |

You got the point about linking carefully a moment ago, but this is a little different. Not only should you link carefully - you also need to link *naturally*.

Again, choose your links well. Forget *most* reciprocal links (unless the return link is from an "authority site" (big company, gov, edu, etc.) And skip obvious black hat link building tactics.

Duplicate Content

This one is based on two of the lesser evils of those seeking quick rankings: laziness and impatience.

Building high-quality pages, loaded with unique content, takes time. That's why it is considered so valuable to search engines. Unfortunately, some webmasters looking for a quick fix decided to take the few great pages they had on their site and duplicate them across the web as dozens – or even hundreds – of mini-sites or doorway pages.

I'll keep this short and sweet. Google hates duplicate content. It's not really classified as "black-hat" but it is considered pretty darned inconvenient. Why, you ask?

Consider how expensive and, well ridiculous, it would be to keep ten copies of every document you have ever written. How much hard disk space would that take? How much slower would your computer run if it was bogged down by nine redundant instances of everything that had ever passed through your keyboard? Now take that and multiply it by five billion. Yes, that's 5,000,000,000. That's what Google keeps track of and they invest a lot of time, effort and money to do it.

How much? Let's just consider the cost of the servers themselves. In 2010, Google had more than one million servers and that number grows constantly. If each cost just one thousand dollars, that's a huge investment. But it's likely several times that!

There is also another reason for Google's disdain of duplicate content. It provides no real additional value to the consumer. Imagine if when researching something via the web every site you went to gave you the exact same information. Everything was described exactly the same and each website offered no more or less information than another. Not a very good consumer experience, right?

2.0	Be Original	Classic Rule

Everything you post on your website is a representation of your company. If it's worth posting, it's worth spending the time to be original.

Think about it... If a search engine can detect duplicate content in a nanosecond, how long will it be before your visitors notice you are just rehashing old stuff - or worse yet, stealing content from other sites? Say goodbye to your credibility.

Doorway Pages, Spinners, and Other Shortcuts

In or around 2004, it became apparent that Google was serious about not wanting to see duplicate content so people who wanted to put up junk doorway pages started copying their homepage text and running it through a spinner to get "unique content." Unfortunately, "unique" usually meant awful because the spinner was just a 'find and replace' function using anything it could pull from a generic thesaurus. I'll show you what I mean...

Original
We specialize in providing a wide variety of modern net business services such as website design, email management and blogging services.

Post Spin
He/she and I focusing on many diverse recent accomplishment mesh tasks performed like website drawing, email overseeing and blogging tasks.

2.0 Avoid Shortcuts to Dead Ends	Classic Rule

In SEO, you will find that most shortcuts lead to dead ends. Tricks like spinners and junk doorway pages worked for a while but, ironically, to nobody's advantage. The quality produced was so bad that most visitors clicked off before they made it past the first paragraph - let alone to the point where they would become a valuable lead. To quote your grandfather, *"If it's worth doing, it's worth doing right."*

Robot Text

One of the more common, and least desirable, optimization tricks is the use of "robot text". This is when the words on a web page are written for a search engine instead of a human visitor. And, it looks horrible. The following is a real example for the cumbersome keyword "make money Internet..."

```
"Make money Internet! With our make money Internet
secrets you will make money Internet right away.
We have been helping people just like you make
money Internet since 1999."
```

How much of that could you possibly read before you decided to poke out your mind's eye just to forget it? Not much for me, that's for sure.

To combat this "bot text", Google invested a great deal of time and money developing a natural text algorithm (NTA) that is still in use today (although much improved).

Bonus Take-away: here's the same message, slightly less annoying, and more geared toward humans:

```
"Make money on the Internet, immediately, using
our 'inside secrets.' We have successfully helped
more than nine hundred people earn an online
living since 1999 and we look forward to helping
you."
```

The second iteration of this message ranks much higher than the first for the keyword phrase, "make money Internet." Although the keyword density (number of times the keyword is mentioned) is significantly lower, the quality score more than makes up the difference. From here just a few more simple tweaks take it over the top...

> "Make money on the Internet with our 'Internet Money ™' training program. We have successfully helped more than nine hundred people earn a healthy full-time income from the Internet since 1999. We can and will help you succeed."

Of course, the main lesson here is to not use page text written only for search engines. An even greater lesson learned however, is that there is no need to. Not only do sophisticated search engines hate bot text, they love natural text.

With just three minutes' work the quality score went from about 7% to 85% and with a few minutes more work it can easily reach 95% or better.

2.0	Write Conversationally	*Classic Rule*

Google's natural text algorithm (NTA) is so advanced that it can actually "grade" your writing as being conversational versus bot-like. It can also determine what level of education a person would need to have in order to understand it.

Tip: *The ideal education equivalency rating is anywhere from 9th grade to 10th grade public school.*

Moving on...

If anything, the rift between unethical SEO people and search engines got bigger in the past decade (2000 to 2010). Plus it was easy to play innocent when the rules were being made up along the way. Now however, every major search engine has hard and fast regulations in place – all publically displayed. In fact, most have had these in place since 2000 or even before.

Google, Yahoo!, and Bing see black-hat SEO'ers as the enemy – and for good reason. These unethical optimizers cost search engines a fortune every year. Why? That's simple. There are three major search engines in the world today: Google, Yahoo!, and Bing – in that order of popularity. If people get lousy results when they search on one engine, the next is only a click away.

How many times have you searched Google, Yahoo! or Bing for something you needed quick, accurate information on, just to be disappointed with lousy search results? More than a few I'll bet.

This is what every search engine hates. It makes them look incompetent. After all, the average user does not know that somebody went out of their way to fool the search engine. All he or she sees is lousy results.

Think about this... In a little over one year Google went from a novelty home-grown search engine to the most popular and profitable in the world. When it comes to anything on the Internet, change happens, and it happens FAST.

| 2.0 | Be a Good Guest | Classic Rule |

Being listed in any search engine's database is a privilege, NOT a right. Unless you paid the search engine for guaranteed organic inclusion (which most do not offer), you are a guest. Be a good guest so you are never asked to leave.

I have heard "They have no right to ban my site!" from black hatters. Well... Actually they do. Always remember that.

Classic Rules of SEO

B efore we explore the new rules, let's take a look at some of the classics. These are the oldies but goodies that your site needs to be following today.

Feel free to make photocopies of the following rules and keep them on hand for SEO evaluations and campaign planning.

| 1.0 | Don't Borrow Keywords | *Classic Rule* |

Whether they are in your main page text, Meta tags, Titles, etc - Do NOT "borrow" any keywords that you have no legal right to use. This is especially true of competitor's names or brand names that are subject to intellectual property law. This is a BIG deal. You can be sued for using a competitor's name or infringing on their intellectual property.

| 1.0 | Stay Keyword Natural | *Classic Rule* |

"Keyword natural" is a quick way of saying you should use all keywords genuinely, as they would normally appear in conversational writing.

Using any keyword a few times per page, only where they make sense, is the way to go. There is no need to cram more in than are really needed. In this case, less is more.

1.0	Visibility = Transparency	*Classic Rule*

Search engines can tell when you are trying to keep things hidden from your visitors, so don't bother. Be transparent. Be REAL. They see "invisible" text as cheating the system and will punish your site if they catch you at it. And don't think for a minute that your visitors are oblivious. This trick is so old that just about everyone knows how to spot it by highlighting big blank areas and watching that hidden text come to life.

1.0	Say it BIG. Say it Loud!	*Classic Rule*

Don't ever use tiny text on a website. It looks awful when people think you are hiding something. And once they figure it out, that is their ONLY area of focus.

If you have something to say, say it big and loud. Even your copyright an disclaimer should be easy to read for anyone who wishes. In fact; this can be a great place for that final keyword.

1.0	Think "Would I Mind?"	*Classic Rule*

Before doing anything that is remotely questionable - take a moment to actually ask the question "If I were in the search engine's place - investing millions to provide the most relevant search results possible - would I mind if someone did that?"

Remember: SEO is about making your site the <u>most relevant</u> it can be <u>for targeted keywords</u>, NOT fooling search engines.

2.0 Updating is Not Replacing *Classic Rule*

Advanced search engines like Google love fresh content but hate page swapping. ALL major search engines know this is a scam to get long-term high rankings from a temporary page full of "spider bait."

Again, add fresh content and remove stale content, but don't swap out an entire page. Show stability AND improvement.

2.0 Link Naturally *Classic Rule*

You got the point about linking carefully a moment ago, but this is a little different. Not only should you link carefully - you also need to link *naturally*.

Again, choose your links well. Forget *most* reciprocal links (unless the return link is from an "authority site" (big company, gov, edu, etc.) And skip obvious black hat link building tactics.

2.0 Link Carefully *Classic Rule*

One of the most time-consuming parts of search engine optimization is link building. And for good reason; it's the most powerful aspect of SEO as well.

That said; choose your links well. Never use link farms of any type. They are reciprocal junk that will get you busted and possibly banned for life.

2.0 — Be Original — Classic Rule

Everything you post on your website is a representation of your company. If it's worth posting, it's worth spending the time to be original.

Think about it... If a search engine can detect duplicate content in a nanosecond, how long will it be before your visitors notice you are just rehashing old stuff - or worse yet, stealing content from other sites? Say goodbye to your credibility.

2.0 — Avoid Shortcuts to Dead Ends — Classic Rule

In SEO, you will find that most shortcuts lead to dead ends. Tricks like spinners and junk doorway pages worked for a while but, ironically, to nobody's advantage. The quality produced was so bad that most visitors clicked off before they made it past the first paragraph - let alone to the point where they would become a valuable lead. To quote your grandfather, *"If it's worth doing, it's worth doing right."*

2.0 — Write Conversationally — Classic Rule

Google's natural text algorithm (NTA) is so advanced that it can actually "grade" your writing as being conversational versus bot-like. It can also determine what level of education a person would need to have in order to understand it.

Tip: *The ideal education equivalency rating is anywhere from 9th grade to 10th grade public school.*

2.0   Be a Good Guest   *Classic Rule*

Being listed in any search engine's database is a privilege, NOT a right. Unless you paid the search engine for guaranteed organic inclusion (which most do not offer), you are a guest. Be a good guest so you are never asked to leave.

I have heard "They have no right to ban my site!" from black hatters. Well... Actually they do. Always remember that.

An Added Edge

While the vast majority of the **_new rules_** require nothing more than awareness and a degree of planning on your part, there are some areas that may require an additional investment in time, money or both. Whenever any additional resources are likely to be needed, you will get an early heads-up as follows:

 **Please Note:** The rule included in this chapter is supplemental and included only to take your SEO efforts over the top. Unlike the other rules, this may require additional time/money/effort.

Even when you see this note, I typically provide a no-cost alternative. Time and effort however, will almost always be considerations. And we all know that's not free.

In short, these sections provide an added edge that can catapult you over your competition in the long-term.

SEO 3.0: The New Rules

The closing months of 2010 saw a great deal of change in the world of search. There are now a whole new set of rules to follow. But these are not the search engine induced policies of yesterday. You know those. The new rules of SEO are insights you need to understand and actions you need to take in order to find success in this new era of search engine optimization; SEO 3.0.

The following pages provide an executive summary of what it takes to earn the top search engine spots - and stay there.

Unlike the eras of SEO 1.0 and 2.0, which typically saw years for major changes to evolve, SEO 3.0 begins at the crux of the greatest industry changes in more than a decade, including:

- Microsoft's Bing working hand-in-hand with Yahoo!
- Google reversing a decade-old *global* strategy to become THE *local* search engine of choice
- Google changing the way the world searches by anticipating your wishes, and providing more options, at the speed of thought

And that's really just the tip of the iceberg. A great deal has changed very recently and I want to put you ahead of the curve.

That said; let's kick this off with a look at the first rule of SEO 3.0, which has to do with protecting your site as it stands today...

SE⊙ 3.0   Remove Your Black Hat   *Rule #1*

Before you do anything else, have a knowledgeable webmaster or SEO person go through your site and get rid of any black hat techniques that might be left over from the old days. This is critical to your SEO success.

Pre-Press Update: *As this book goes to press (January 25th), Google has just announced a new initiative to crack down on all SEO spam techniques. **Check your site before it's too late.***

The Changing Face of SEO

Search engine optimization is always changing and evolving. It's natural. And as with the development and progression of all things, it's inevitable.

Why all the changes? One word… Competition. The search engines continually transform themselves in an ongoing effort to improve their positions and popularity as the engine of choice. After all, this is a multi-billion dollar industry we're talking about. What's the drawback to shaking up the industry a bit if it adds a digit or two to the end of their year-end numbers? As far as they're concerned, there is no downside. It's all upside.

And of course, when the search engines make changes, the rules and strategies of search engine optimization change right along with them.

In the past few months alone, Google has once again made changes to its core functionality through the introduction of "Google Places", "Google Instant", and "Place Search", while Yahoo! now displays Bing's search results.

Until recently, most of the constant changes, made by search engines, have been small. A minor algorithm update here. A new "sponsored listings" mod there. It was all relatively small stuff.

That was then.

In the last quarter of 2010, four HUGE search engine game-changing events happened mere weeks apart and the fate of billions of web pages changed instantly.

That's evolution. And as with all things that evolve, you can adapt or get left behind. This book was written to make sure you don't get left behind. As a business owner, half the battle will be understanding that in the world of search engine optimization change is inevitable. Hence, SEO 3.0 Rule #2 …

SE◎	Welcome Change	Rule #2

Merely accepting change is no longer enough. If you want to succeed in the fast-paced world of search, you need to embrace it. Once you can anticipate change, based on experience and newfound knowledge, you are in a position to take the lead when opportunity arises.

Just remember that change can lead to self-fulfilling prophecy. If you expect problems, you will most certainly find them.

How Are You Evolving?

How did your search engine optimization (SEO) strategies change when Google, Yahoo!, and Bing altered the game? Or, did they change at all?

If you ask most website owners about the three biggest search engine occurrences in over a decade, they will say something like "Hmm, I thought Google looked different," or, "I noticed my Yahoo! rankings changed," but they don't have much more to say on the subject. And that's fine – as long as they have a solid Online Marketing or SEO person taking care of the details. When an expert is on top of things there is no reason to sweat the small stuff, even when the "small stuff" is kind of a big deal.

But, what if no one is minding the store? How will your online business be impacted? If you have not taken any time to consider how to handle these and other world-changing events, now is the time to start. Or, hire a professional to do it for you. Just make sure they keep you in the loop.

Remember, search engines change all the time. But now a decade's worth of changes can creep up in a matter of months. And, years of experience has shown that those who are prepared ride the top of that success wave, while those who are not get crushed in the breaks.

SE◉ 3.0 Evolve or Get Left Behind *Rule #3*

In SEO, or any online marketing, there is a cold, hard fact that must be faced... ***Those who do not evolve, get left behind.***

The web is not like print advertising or direct mail where yesterday's techniques will work for years - or even decades. Everything online happens FAST. You need to evolve *quickly* just to keep up and *willingly* - even <u>eagerly</u> - to get ahead.

Forget Clicks - Get Leads

S EO 1.0, in general, was all about getting clicks – any clicks. SEO 2.0 was the age of getting more highly targeted clicks. Better, but still not as good as it could be.

We have grown past the age of "clicks". Now is the time for getting LEADS. Clicks don't pay the bills. Leads have the potential to pay-out ongoing dividends with every conversion of lead to sale - and then of course, future sales.

With this in mind, NetSearch Direct did something unheard of in 2006. We told clients we would provide valid leads from their SEO, instead of just clicks.

In the minds of many traditional SEO companies, this was crazy. To them search engine optimization was all about the process of obtaining page one organic rankings on a search engine results page (SERP). What happened after that was, well.. it wasn't SEO.

Okay, it was not SEO, but it was marketing and that was my background. To me, page one rankings were just the first step in the process of giving business owners the fuel that energizes a business…new customers.

Taking search engine optimization to this level requires a true partnership between your SEO company and your business. It may mean that you will have to give up some control over the look and feel of your site. If your SEO company's job is to produce leads then you must allow them to follow the visitor throughout the site and track what leads to conversions. We found that most savvy business people with a need to succeed are likely be open to hearing suggestions about site changes that would drive sales. ***Providing leads over clicks was a gamble that paid off in spades – both for our clients and for us.***

Getting genuine leads changes the game. It eliminates half the legwork of the sales process and has an immediate impact on your bottom line. Leads can be instantly monetized – and should be, without question. The time has come to demand more from your SEO efforts. Forget clicks. Get leads. And if a modern day SEO company tells you they are not willing to stand behind their relationship with you, and make sure you get qualified leads, then run away and don't look back.

To really succeed, you need a partner, not a parasite. And an SEO professional who is just in it for the "clicks" is of little help.

SE⊚ Forget Clicks. Get Leads!	Rule #4

Clicks are worthless unless they result in leads or sales. For most companies offering a product or service over $250, the goal is to get a qualified lead in order to close the sale - with the understanding that a qualified lead is the halfway point to a sale. That said; cut your sales legwork in half by focusing on leads over clicks. The "how to" of it could fill an entire book but simply asking your next SEO person "How can you get me leads instead of just clicks?" only takes a moment.

The Big Three Are Now the Big Two

Although technically, Yahoo! is still Yahoo!, the reality is it has lost its independence. It has been consumed by Microsoft's Bing search engine and now shows virtually – if not exactly - the same search results as Bing. So, while you have been told for years to optimize for Google, Yahoo!, and Bing, now the optimization efforts – for the foreseeable future at least – should go to Google and Bing. Yahoo! is just along for the ride, happily displaying Bing's ranking results (typically in a *similar but not exact* ranking position match.).

"So what's the big deal?" you might ask. Consider this… Yahoo! has twice the market share of Bing. That said, chances are if you ever had your website optimized, most of that effort went into Google and Yahoo! with *maybe* a few scraps of attention left over for Bing.

All the effort that went into Yahoo! is now…what? Well, I wouldn't say wasted, but I would say potentially minimized.

By potentially minimized I mean that if your SEO team was on the ball, you have little to fear. You can turn this around in a matter of weeks and your losses would have been minimal to begin with. But, what if this is not the case?

Ask yourself, "Can my business survive losing all of that Yahoo! traffic?" And, "What will the damage be to my bottom line?"

By the time you read this, you probably already know the answers to both of these questions. The main purpose of this section is to get you thinking ahead to future changes and how to minimize the impact to your online business, as well as to keep you in the loop about what happened and when.

It's hard to plan for the future when your knowledge of what went wrong in the past is limited – especially in the area of search engine optimization, which is always changing.

Now, about the Big 3 – or rather, Big 2…

As you can see, Yahoo! accounts for a big part of the search market. And although, the data supplied is all Bing, the results per search engine are not always exact.*

*Example: The same keyword may be #1 on Yahoo! but #4 on Bing, etc.

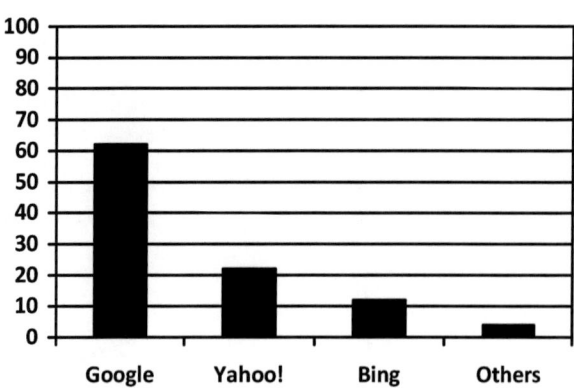

The Big 3 - *Search Engine breakdown by popularity*

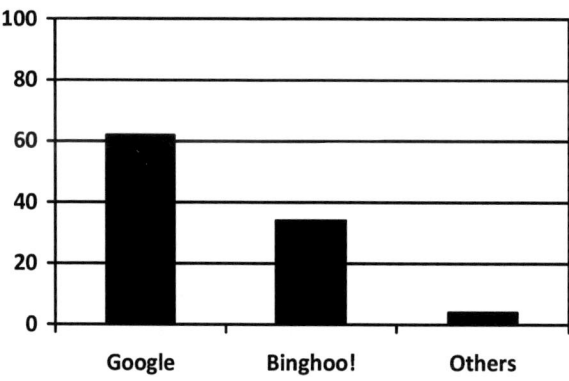

The Big 2 - *Gives you an idea of Yahoo! and Bing combined*

Knowing really is half the battle. The other half is being the unique, prepared business professional who does not panic at every change and is able to adapt with ease. This is what makes all the difference.

But I digress... There are three main takeaways here, as follow:

First: Things can change quickly so we need to be prepared for that change (more on that later). As important however, we need to not panic when change occurs.

Second: It's foolish to place all of your eggs in one basket, unless that's the only basket you care about. Bing was always important and this could just as easily have gone the other way.

Third: Well...We've got a rule for that...

SE🌐3.0 Make Bing a Priority Rule #5

Bing is no longer a *"nice to have"*. It is a **must have** unless you want to completely ignore one-third of your potential visitors.

Their deal with Yahoo! is big news but more importantly is the reason behind the deal. Microsoft (Bing) sees search as something they want to be in - in a very BIG way. How often do they lose when they really want something?

Can Google Read Your Mind?

O n April 1st, 2000, Google played one of their original and still legendary April Fool's Day pranks, about a technology called MentalPlex. Take a look at this:

"MentalPlex is the only search engine that accurately returns results without requiring you enter a query," a FAQ created especially for the joke explained. "Google's CEO and co-founder Larry Page calls MentalPlex 'a quantum leap in finding what you are looking for on the Internet.' Typing in queries is so 1999." – PC World

As far as jokes go, this was a good one. A lot of people fell for it. But, now the joke is on everyone who scoffed at the idea of Google reading your mind because they are currently about as close as any technology is likely to get (hopefully).

Google Instant, one of their latest and greatest, is a natural expansion of an earlier experiment in which Google tried to anticipate your search needs based on the first few letters you typed into the search box.

Here's how it works...

Type in "red" and you would see possible auto-complete words ranging from "red box" to "red lobster menu." This was basically just a simple auto-complete function, but Google Instant is really a different animal all together.

Now, type in "red" and you will see similar options but not just the words. You will also get updated pages with full results for whatever "guess" is at the top of the list. So *instantly* Google changes the organic search listings to match the phrase at the top of the search box. And, did I mention Google knows where you are searching from based on your IP address (Internet Protocol or computer address)?

web   Images   Videos   Maps   News   Shopping   Mail   more

Google

red
red **box**
red **lobster**
red **robin**
red **door spa**
red **cross**
red **wing shoes**
red **redemption**
red **versus blue**
red **dragon tattoo richmond**
red **lobster menu**

Google Instant trying to anticipate your every search

In this example Google Instant is showing results on the name "red box," which is basically a DVD rental vending machine in Richmond, Virginia – the area I am searching from.

With Google Instant, the most popular search terms for any particular area of interest are now being suggested to your would-be customer as Google tries to read their minds and anticipate their wishes.

OK, that sounds like the old auto-complete feature. But there's more to it…

Now, Google does not wait for you to select the keyword you want, or even to stop typing. It simply refreshes the page with completely new search results and paid listings every couple of seconds (or less) depending on what you type - as you type.

So what's the rationale behind it? I'll let Google explain it in their own words (taken from Google.com/instant/):

"Google Instant is a new search enhancement that shows results as you type. We are pushing the limits of our technology and infrastructure to help you get better search results, faster. Our key technical insight was that people type slowly, but read quickly, typically taking 300 milliseconds between keystrokes, but only 30 milliseconds (a tenth of the time!) to glance at another part of the page. This means that you can scan a results page while you type.

The most obvious change is that you get to the right content much faster than before because you don't have to finish typing your full search term, or even press "search." Another shift is that seeing results as you type helps you formulate a better search term by providing instant feedback. You can now adapt your search on the fly until the results match exactly what you want. In time, we may wonder how search ever worked in any other way.

Benefits

Faster Searches: By predicting your search and showing results before you finish typing Google Instant can save 2-5 seconds per search.

Smarter Predictions: Even when you don't know exactly what you're looking for, predictions help guide your search. The top prediction is shown in grey text directly in the search box, so you can stop typing as soon as you see what you need.

Instant Results: Start typing and results appear right before your eyes. Until now, you had to type a full search term, hit return, and hope for the right results. Now results appear instantly as you type, helping you see where you're headed, every step of the way."

But wait, there's more. Read on to see how this new feature can save you time while increasing efficiency.

"Did you know:

Before Google Instant, the typical searcher took more than 9 seconds to enter a search term, and we saw many examples of searches that took 30-90 seconds to type.

Using Google Instant can save 2-5 seconds per search.

```
If everyone uses Google Instant globally, we
estimate this will save more than 3.5 billion
seconds a day. That's 11 hours saved every second.

15 new technologies contribute to Google Instant
functionality."
```

Again, this is pretty big stuff. Google is always trying to change the game and this is a pretty big change. In fact, if your website is not optimized for the <u>most popular</u> search terms (keywords) people enter into Google, your traffic might slow to a crawl.

After all, Google is not even asking people to finish their own thoughts. They are putting potentially hundreds of options in front of the searcher before he can even finish typing out what he was originally looking for.

Give it a try yourself. Did you finish typing the keyword phrase you had in mind?

And, given Google's track record, it is easy to see they do not typically go backward. This means that this change – or some variation of it – is likely here to stay. But, what else does it mean to you?

Let's reverse the scenario and try to read Google's mind for just a moment. Remember the new Google is intuitive. It *instantly* delivers you search term options as you are typing. It *remembers* your particular search history and it takes into consideration from what *geography* you are searching. If you controlled the most popular, powerful, and profitable search engine on the planet what would your next move be? Chances are it would be something to even further define its success in dealing with *local* search.

In terms of evolutionary scale, if Google's old auto-complete search feature was a leap from 1 to 3, this new initiative is 5 to 10.

Don't make the mistake of under estimating the importance of selecting the right keywords for optimization. Google Instant has made this more important than ever, as it is automatically suggesting the top keyword phrases to the searcher. Make sure your SEO firm puts a premium on researching the most searched terms for your business category. And definitely get rid of the junk keywords.

SE◉	Get Rid of Your Junk	Rule #6

Junk keywords cost you time and money, so get rid of them. Now this does not mean highly targeted, yet low volume keywords with potential. This means absolute junk that nobody is searching for. So how can you tell? I subscribe to several services, but Google offers a nice freebie at:

https://adwords.google.com/select/KeywordToolExternal

Getting Local With Google Places

Think Google Places is just some convenient little map that pops up when you search for a product or service available in your area? Think again. It's all that times 10. Just read Google's press release of April 20, 2010 to see what I mean. But first, I should tell you that although only a few months old, this information is already *almost* obsolete as this was merely the first step in a much bigger plan.

I stress the word "almost" because understanding how Google's latest search strategies evolved is going to paint a picture worth a *million* words when it comes to anticipating their next move(s).

Here's what Google Places looked like, followed by the press release that gives you insight to how they think:

Local business result for **pizza** near **New Haven,CT**

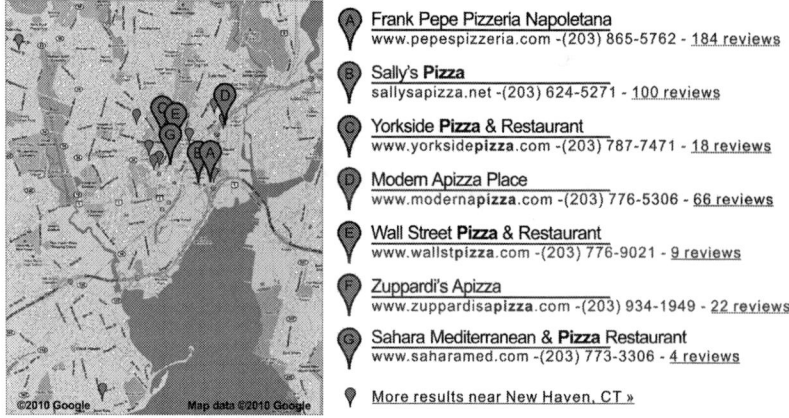

Google Places' famous "7-Pack" (this is where it all started)

Announcement

April 20, 2010

Google Goes Local with Google Places

Every day, Google connects millions of people with businesses in their local communities. We help people find these businesses when they search Google and discover Place Pages, and we help business owners manage their Place Pages on Google through a tool we call the Local Business Center.

Today, we're renaming the Local Business Center to Google Places and adding a number of new features. In addition to these new features, Google Places will continue to offer the same tools as the Local Business Center, such as helping a company verify and supplement business information including hours of operation, photos, videos, coupons and product information; providing a way to communicate with customers; and giving businesses new insights that enable it to make smart decisions.

The new Google Places name simplifies the connection with Place Pages and reflects our ongoing commitment to providing business owners with powerful, yet easy-to-use tools to help people discover them when they search.

"Getting customers through Google makes my job great. It's incredibly valuable to have all those listings grouped together in one place," according to Chris Gallagher, owner of Bean & Leaf in New London, Connecticut. "It shows you how important

Google Places is because we want to truly reflect the attention we pay to our business to our customers."

Today we're introducing several new features to Google Places:

Service areas: If you travel to serve customers - a photographer, plumber, or piano tuner, for example - you can now show which geographic areas you serve. And if you run a business without a storefront or office location, you can now make your address private. This helps the millions of home-based and service businesses be found by customers in their service areas.

A new, simple way to advertise: For just $25 per month, businesses in select cities can make their listings stand out on Google.com and Google Maps with Tags. As of today, we're rolling out Tags to three new cities — Austin, Atlanta and Washington, D.C. — in addition to ongoing availability in Houston and San Jose, CA. In the coming weeks we'll also be introducing Tags in Chicago, San Diego, Seattle, Boulder, and San Francisco.

Business photo shoots: In addition to uploading your own photos, businesses in select cities can now request a free photo shoot of the interior of their business which we'll use to supplement existing photos of businesses on Place Pages. We've been experimenting with this over the past few months, and now have created a site for businesses to learn more and express their interest in participating.

Customized QR codes: From the dashboard page of Google Places, businesses in the U.S. can download a QR code that's unique to their business. QR codes can be placed on business cards or other marketing materials, and customers can scan them with certain smartphones to be taken directly to the mobile version of the Place Page for that business.

Favorite Places: We're doing a second round of our Favorite Places program, and are mailing window decals to 50,000 additional businesses around the US. These decals include a QR code that can be scanned with a smartphone to directly view the mobile Place Page for the business to learn more about their great offerings.

Over the past few months we've also added the ability for business owners to post real-time updates to their Place Page. They can promote sales, special events, or anything else customers need to know right now, and this feature lets businesses communicate that directly to their customers. They can also provide extra incentive by adding coupons, including ones formatted for mobile phones.

To keep track of how a business listing is performing on Google, we offer a personalized dashboard within Google Places that includes data about how many times people have found your business on Google, what keywords they used to find it and even what areas people traveled from to visit their business.

With the dashboard, a business can see how the use of any of these new features affects interest in the business and can make more informed decisions about how to be found on Google or how to interact with your customers.

One out of five searches on Google are related to a user's location, and very often people are looking for local businesses. Google Places is just the beginning of what's to come from our efforts to make Google more local. To learn more, you can visit our blog post on Google Places or see our newly updated Help Center. We'll also be posting on the Lat Long blog throughout the week to give a deeper dive into many of our newest features. To get started now, go to google.com/places.

END

As you can see, Google is doing more than just adding a bunch of virtual pins to a map; it's redefining how people find what they want online at a local level. And, if your store can be part of the famous 7-Pack (the seven local listings shown in Google Places) you can expect to see an increase in visits.

In fact, with the introduction of Place Search, you don't even need to worry about the 7-Pack because more often than not these listings will be showing throughout the regular organic listings.

In short, being found in Google Places, whether in the regular organic mix or in the 7-Pack, is an incredibly important factor for any local business with a physical address. Do whatever it takes to get there and stay there.

SE◉	Get in Google Places	Rule #7

Google Places is <u>going places</u> and your website needs to be along for the ride. If you have never looked into it, and claimed your free business directory listing, now is the time. It takes a phone call, possibly a postcard, and some patience, but is so worth it. As you are about the see; this is one of your most important "to do's". Visit at http://www.google.com/places/ or have your SEO person do it for you.

Google Place Search

What about part two of Google's bid to be the local search solution, "Place Search"? Never heard of it? That's okay. I can almost guarantee you that you have seen it – and possibly even used it without realizing it.

With Place Search, Google is going a step further in the march toward becoming THE *local search engine*. This is accomplished by simply integrating the Google Places info into the regular search results. It's really pretty brilliant. They shake up the industry and appear to present entirely new rankings, using their old data. This, of course, ends up trumping some long-standing top placers and it all goes back to good old Google Places (previous chapter).

---

nyc pizza	Search

About 2,990,000 results (0.25 seconds)   Advanced search

Best Pizza in New York City - Great **Pizza** in **New York City** ☆
When you visit **New York City** you don't want to miss out on the delicious **pizza**. Whether you want a whole pie or just a slice of **pizza**, **New York City** is a ...
gonyc.about.com/od/restaurants/tp/best_pizza.htm - Cached - Similar

Top 10 Ten Best **Pizzas** in **New York City** Restaurants **NYC** New York ... ☆
This restaurant is one of the best **pizza** parlors in **New York City**.
✚ Show map of 328 E 14th Street, New York, NY 10003
www.gayot.com/.../best-newyork-ny-top10-pizza_1ny.html - Cached - Similar

Lombardi's Pizza ☆
New York City was the birth place of New York style **pizza**. During the year of 1905, Lombardi's was licensed by the City of New York, becoming America's ...
www.firstpizza.com/ - Cached - Similar

 32 Spring Street, New York - (212) 941-7994   ★★★★☆ 764 reviews
"All in all, a great NY pizza pie. It is worth traveling to this end of   Place page
towm ..." - tripadvisor.com (130)
yelp.com (957) - citysearch.com (227) - menupages.com (110)

John's Pizzeria ☆
Call for Holiday Scheduling; closed for Thansgiving, Easter and Christmas.260 W 44th St, New York, NY 10036, (212) 391-7560 ...
www.johnspizzerianyc.com/ - Cached - Similar

 260 W 44th St, New York - (212) 391-7560   ★★★★☆ 442 reviews
"Pro: Great service, ambience, and bar. Con: mediocre pasta/sauce   Place page
All in all ..." - nyc.com
yelp.com (226) - tripadvisor.com (112) - citysearch.com (102)

This is serious stuff, especially if you have built a national online sales strategy that might now be invisible to sixty percent of your target market because of Google's most recent changes, changes made to provide the most relevant local search results. To obtain page one rankings for local keyword phrases on Google one must integrate a successful Google Places/Place Search strategy into their SEO plan.

SEO 3.0   Place in Place Search   *Rule #8*

Times have changed. The old "Google Places" results are no longer hiding in the corner, hoping to get noticed. Now they DOMINATE the first search results page in a section called **Google Place Search**. Your potential visitors can't miss them so you cannot afford to miss out on being found within them. To do this you must first claim your Google Places Directory Listing in and try to build up some good reviews.

Google Instant Preview

Google Instant Preview is a bit of a milestone. Knowing they cannot always keep the lower quality sites from being found in a top search position, Google has added a feature that allows users to preview what a page looks like before committing to a click. This feature could be a game-changer – *if* it catches on.

It's a pretty simple in concept. Anytime you see the small magnifying glass in the upper right-hand corner of the search result, you can click on it or any white space within that search result area and an instant preview will appear to the right.

---

pizza parlor new york

About 32,700,000 results (0.29 seconds)

New York City **Pizza Parlors** | Pizza Restaurant in **New York** City, NY
Find **New York** City **Pizza** on Magic Yellow. Yellow Pages online for **Pizza** in **New York** City, NY.
www.magicyellow.com/category/**Pizza/New_York**_City_NY.html - Cached

Best **Pizza** in **New York** City - Great **Pizza** in **New York** City
When you visit **New York** City you don't want to miss out on the delicious **pizza** . Whether you want a whole pie or just a slice of pizza, New York City is a ...
gonyc.about.com/od/restaurants/tp/best_**pizza**.htm - Cached - Similar

Top 10 Ten Best **Pizzas** in **New York City** Restaurants NYC New York ...
This restaurant is one of the best **pizza** parlors in **New York City**.

Google Instant Preview close-up

In the above example I have clicked on the magnifying glass of the middle listing. This is just a close-up so you can see the magnifying glass. The following is the entire screen…

SEO 3.0

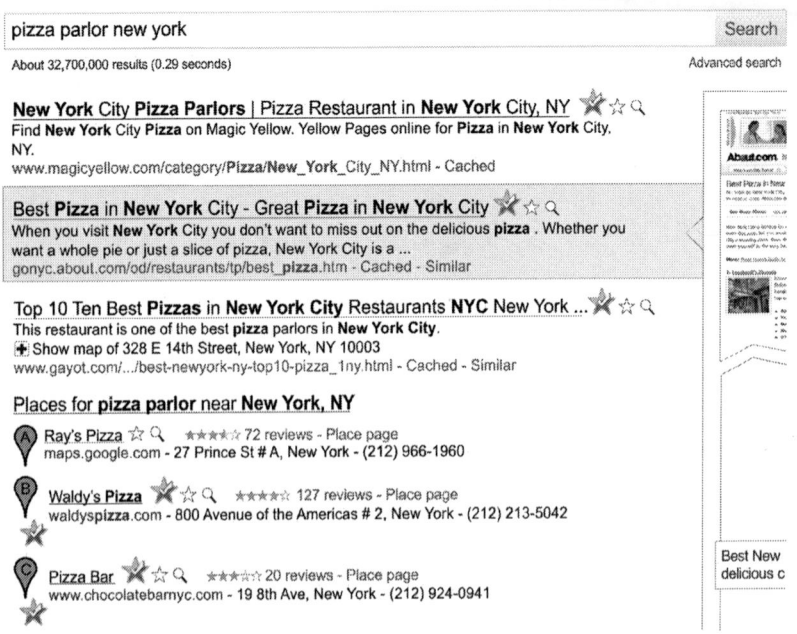

Google Instant Preview as it appears to the user in full screen

Here is how Google describes the benefits of this new feature:

Quickly Compare Results
A visual comparison of search results helps you pick which result is right for you.

Pinpoint Relevant Content
Text call outs highlight where your search term appears on the web page so you can evaluate if it's what you're looking for.

Interact with the Results Page
Page previews let you see the layout of a web page before clicking the search result.

X

The "X" close-out option in the upper right corner, not only shuts off this particular preview but the actual preview feature itself - until you re-enable it by clicking on one of the magnifying glass icons in the search listings.

A jagged break indicates the page is longer than what is being shown. If there is content of greater relevant interest at the bottom of the page, the break will appear in the middle, omitting the center section. If the more relevant content is in the center area, the break will appear at the bottom.

This is a special call-out to draw attention to something of interest. It might be anywhere.

After reading the benefits of Google Instant Preview and actually seeing it for yourself, you can likely understand how this could change the way people choose to visit web pages. Now, they can just preview each page and see what it looks like.

And, if you are thinking "who cares?" I've got bad news for you: people really do judge a book by its cover and they most certainly judge a website by appearance - within two seconds of a screen-shot scan!

By the way, if Google Instant Preview looks familiar it's because Bing did something very similar more than a year ago. Go to Bing.com and check it out. It's still in place today.

You can think of Google Instant Preview as a 2.0 version of the technology. It eliminates the need to "commit to the click."

For you this simply means your preview needs to impress searchers as much as your keywords impress search engines.

SE⊚ 3.0 Take The 2 Second Test	Rule #9
Now that people no longer need to visit your website to see if they like it, you need to look at it in a whole new way - with fresh eyes, and for no longer than 2 seconds. That's how much time the average person will spend looking at it before they commit to the click. So print out some color screenshots of your home-page and a few of your top online competitors and honestly choose which is best - with just 2 seconds to decide.	

Keeping Track with Google Analytics

I f you managed a website back in the late 1990's you probably remember having two choices in analytics: pay for an app like "*WebTrends*" or simply guess how people were finding your products and services online. That all changed with Google Analytics in late 2005 (which evolved from "Urchin").

With Google Analytics, anyone could get world-class analytical data about their website for free. And, it is so much easier to use than anything before. All you have to do is sign up for a free account, enter in your URL, and paste a little snippet of Java-Script code into each web page's activity you wanted tracked.

Talk about cheap and easy! I remember paying a small fortune and jumping through virtual flaming hoops to get half that much useful data just a few short years before.

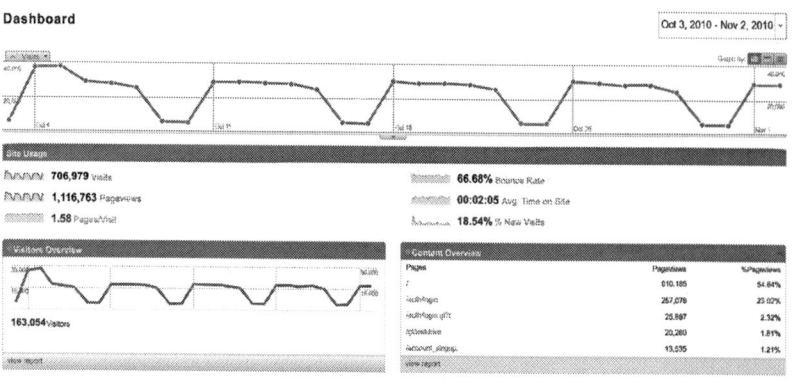

Google Analytics Dashboard

So what can Google Analytics tell you about your visitors? More than most people ever dream!

Of course, there is the regular stuff like:

- Seeing the **exact keywords** people use to find your site
- Knowing **what pages** visitors enter and leave from
- Having a virtual stopwatch tell you **average time on site**
- Understanding **what pages are most popular** and when
- Discovering which search engines are driving the most – or the **most qualified** traffic
- Picking out the referring **web pages** to see where your traffic is coming from – other than search engines

But there is so much more. You can also...

- **Set goals** to track every time an action is taken, such as a purchase made, form submitted, etc.
- **Create funnels** to see where people drop off during multi-part form fills or purchases
- **Exclude traffic stats** from any undesired source...so you don't count your own hits, for example
- Learn what **browsers visitors use most** so you can use your web design budget wisely
- **Automate reporting** and have the results emailed
- Setup conversion goals then check and **track the results by geography**

This is just the beginning. Far too many people overlook what Google Analytics can do for their business so maybe this will help. Six or seven years ago most companies had to pay upward of one thousand dollars (or even more) per year to get just half of this data. And, if you used a marketing company to track your stats and offer reports not nearly as in-depth as those offered by Google Analytics for free, you likely paid several hundred dollars per report.

But be warned, Google Analytics is not perfect. It's absolutely amazing for a free tool that does not pour through your server logs, but as such it has some small drawbacks.

For example, it uses "first party" cookies (small bits of code it saves to the user's computer), which are the least intrusive kind, but still cookies none the less. This means that while Analytics will pick up on probably ninety-eight percent of your traffic or more, there is that small segment that declines all cookies. Hence you won't get accurate tracking on that segment.

A bit inconvenient? Yes, but not really a big deal. The great news is that your site will work just fine for these people as well as those who accept cookies. You just won't have accurate tracking for that tiny group of visitors.

To summarize, not taking advantage of Google Analytics is absolutely insane. If nothing else, you should install the basic script with no special goals or funnels and let the data collect.

Without the customizations you won't have the in-depth report-ing that makes this tool better than most paid versions, but you will have the basic data that any SEO company, marketing expert, or web specialist will need down the road when you want to get more business.

A Note on Cookies... Cookies are just small text files stored on your computer. Some of these, called "session cookies", will disappear once you close out your browser, but most will stay in your hard drive. They are simply there to identify you to the web page you are visiting to eliminate the need for you to re-select user preferences, etc.

Like all good cookies, they come in different flavors... First-party cookies, the kind that Google Analytics uses (and the least intrusive) are bits of text set by the same domain that is currently in your browser's address bar. Third-party cookies are considered more intrusive, and therefore are more commonly blocked.

SE**⊚**3.0	Use Analytics	Rule #10

Having the information provided by a good website analytics program is crucial to long-term SEO planning and success. This can be a free app like Google Analytics (GA) or a paid app such as WebTrends. Just select one that provides, at a *minimum*, traffic stats, keyword tracking, goal tracking, and the ability to get custom reports at the push of a button. There are plenty on the market, with Google Analytics being the leader.

Goals and Conversions

We briefly mentioned goals and conversions when we looked at Google Analytics, but the topic deserves some more attention, so let's dig in.

Goal and conversion tracking is the info you would have paid hand over fist for just a few years ago, and Google provides it for free. So, why is this data so valuable? Because it lets you create your own repeatable process as a future shortcut to success – in whatever area you wish.

Looking for form fills? Goals and conversion data can show you not only that a form fill was completed, but whether it came from a paid source (like PPC), organic search, or if it was a referral from a banner ad, blog post, or article. That's powerful stuff. Once you have insight to this previously hidden world of information, the online world is your oyster.

Let's check out an example… Say you have three pay-per-click (PPC) ads running and each leads to a different landing page containing a unique offer. Ad 1 goes to landing page 1, or LP1, Ad 2 goes to LP2, etc. In a matter of minutes, you can enter each landing page URL as a goal to be tracked as well as each thank you page (the visitor hits after completing the form fill or other call to action). Since the user will only visit that page after completing the call to action, this can be considered a conversion or simply another goal, as you wish.

If Ad 1 is based on a keyword that costs $2 per click, and it typically takes fifty clicks before someone actually completes the call to action to become a solid conversion, you now know it takes $100 per conversion – whether that "conversion" is a form fill, sale, or whatever.

How about Ad 2? Thanks to Analytics we can see that the keyword for that ad costs an average of $2.50 and takes about sixty clicks to get a conversion; so it is costing us $2.50 x 60 = $150 per conversion. Wow! With that little bit of change tacked onto the cost per click and a few more clicks it really ads up fast.

Let's look at Ad 3. It costs $1.75 per click and it typically takes forty clicks to convert. A little quick math tells us that $1.75 x 40 = $70 – or less than half of what Ad 2 costs per conversion.

We just multiply the Cost Per Click by the number of Clicks To Convert, to get the Cost Per Transaction (or per lead).

ID #	CPC	CTC	CPT
1	$2.00	50	$100.00
2	$2.50	60	$150.00
3	$1.75	40	$70.00

ID# = Ad Identification Number (1-3)

CPC = Cost Per Click

CTC = Clicks To Convert

CPT = Cost Per Transaction

Now you know exactly where your PPC ad budget is best spent. You can even take it a step further and swap out landing pages with different keywords to find the ultimate best deal.

Incidentally, these are real numbers taken from a real account, rounded for simplicity sake in this example. This is not guess-work. It's historic data that a good SEM company will use regularly to tweak your ad spend and maximize your return on investment.

What was the end result? In this particular case we were able to get this client down to $1.63 per click with 38 clicks needed on average per conversion for a cost of just under $62 per conversion. When they came to us, they were spending $148 per conversion on average.

Ad ID	CPC	CTC	CPT
1	$2.00	50	$100.00
2	$2.50	60	$150.00
3	$1.75	40	$70.00
4	$1.63	38	$61.94

Does this sound crazy expensive? Here's the *really* crazy part. At $148 per conversion they were still making money so they never questioned it. Now they make a TON of additional bottom line profit and continually reinvest some of the saved revenue into testing new keywords and campaigns. That's smart business. Not all campaigns are winners like this, but their lowest producing campaign is still less than $95 per conversion on a $1,600 sale.

Just think about it. It takes only minutes to set up a goal in Google Analytics and doing so can help you increase profits exponentially. Where is the downside?

SEO **Set Goals & Track Results** *Rule #11*

Using Google Analytics, setting goals and tracking results is insanely easy - and free! Do it and reap immediate rewards.

Advanced Tip: If you have a multi-part form (such as a 2 or 3-step sign-up process) you can quickly add a funnel to learn just about everything you ever wanted to know about where conversions come from, and where they go to die.

Get Cracking on Call Tracking

Spending hundreds or even thousands of dollars setting up multiple telephone lines and assigning different ad codes to each, then piling through dozens of sheets of call logs is what most people envision when they think of call tracking. After all, that's how it was done a decade ago and beyond. But, today it's a whole lot simpler and cheaper.

A good SEM or marketing company can take care of all the complicated bits and pieces and just let you enjoy the benefit of knowing what calls came in from where – and what was said, if that's of interest.

Here's what you should expect:

Expect to pay $25 to $40 per month, per additional phone line initiated for call tracking purposes and possibly a "per minute" rate of 10 or so cents. One new unique call tracking number per voice line (whose number is on your website) is usually enough. And remember, this is on inbound calls only, so your per minute costs can easily be minimized.

Expect to provide a "forward to" phone number. This can be any telephone line or even a cell phone.

Expect to be asked to record a short message or have the company do it for you – usually at no additional charge.

Expect to be given the option to have the inbound calls recorded "for quality and training purposes" (with caller notification of course).

Expect to receive a call tracking console login of your own so you can check your stats or even listen to recorded calls anytime you wish.

Expect to receive a regular report that provides all of your call stats in an easy to understand manner. This is typically provided on a monthly basis.

And finally, expect to get more useful information out of this call tracking information than you ever thought possible.

With your very first report you will be able to spot calls that dropped off in less than one minute, and even those who just hung up. You will be able to listen in on calls to see how your employees treat customers and how customers treat employees. No more "he said, she said" headaches!

Here's how it Works:

A call tracking phone number is secured on your behalf and entered into the call tracking system. Once in the system, a small snippet of JavaScript code is created that works just like the "find and replace" option of a word processor like Microsoft Word. Each time a person finds your website through a designated search engine organically, the phone number on your site is updated to show the call tracking telephone number. The searcher dials that number and activates the call tracking system, which forwards the call to the line of your choosing, which means...

1. That particular phone number is called only when the person searching found it on your website, and...
2. Only if it was found by clicking on an *organic* search match that led them to that page.

The rare exception is when someone writes down that (script generated) phone number to call later. The effect however, is the same. That person found you initially through an organic search and you can easily track that fact, whether it's now or later.

<u>Very Important</u>

Don't be fooled by some SEO companies who claim to provide call tracking of the organic search results when in reality they are actually hard coding a unique number on your website. The only thing this does is tell you that the caller found that number on your website; however, in this case you have no idea how they *arrived* at your site. They could have been referred from another site or PPC campaign or typed your URL directly into their address bar – none of which would be the result of search engine optimization. As a business owner it is extremely important to know which marketing dollars are producing results (calls in this case) and which are not.

Many of our customers also purchase additional unique phone numbers which they can utilize in other marketing vehicles (yellow pages, direct mail, newspaper, radio, etc.). If they are all feeding into one call tracking console you will have one place where you can easily review the effectiveness of all your marketing campaigns. Now that's powerful!

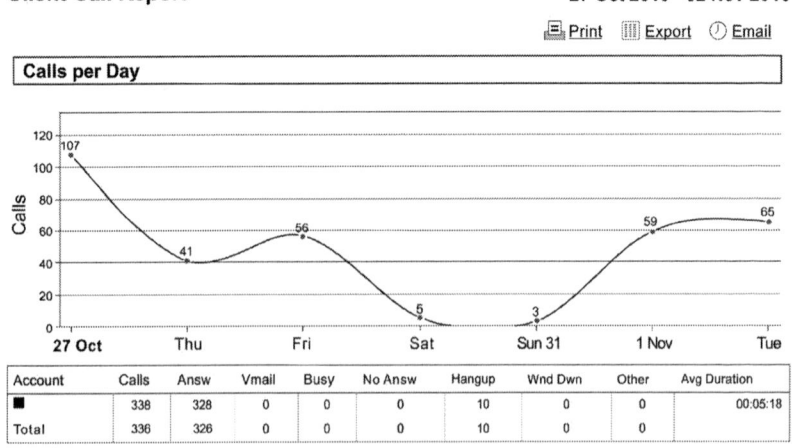

Call Tracking Console – One of Many View Options

SE⊚ 3.0   Track Your Calls   Rule #12

If the business you receive via phone calls is at all important, invest in call tracking. It's worth the extra $40 or so per month. In fact, ***it's too cheap not to use***. When properly set up it's like having a 24/7 operator showing you exactly how each piece of new business came in. It will even let you pick the search engines you want to track calls from, whether the calls were generated by organic search or ads, etc. This is a BIG deal.

Should You Commit to Submit?

It's important to know that search engine submission is not the same thing as search engine optimization. Although submission is often an important part of your SEO campaign, just telling search engines about your site is usually not enough to get any kind of decent rankings – or resulting traffic – from that engine.

In fact, most websites don't need to be submitted because the search engine spiders (AKA / crawlers and bots) will find them through inbound links. But of course, this means you have to have inbound links from web pages that the search engine visits, which in turn means you might be waiting a while before a spider ever crawls your site.

So is it better to submit your site to the search engines or should you just wait for them to find you? The answer is, an ambiguous, "it depends." It depends on if you have the time to wait for a spider to naturally find your site. It depends on if you believe the rumor that Google gives less credit to sites that were submitted rather than found naturally. It depends on how you plan to submit your site, if you head in that direction. Yes indeed, it depends.

I tend to err on the safe side and not submit sites to Google and to only hand submit sites to Yahoo! and Bing. As for Google, I concentrate on quality inbound links to get the site picked up naturally and quickly by creating links on pages that Google considers "fresh," meaning they come back and spider the content regularly – often daily.

And what was that about hand submitting? Why on earth would I take the time to manually submit a website to a search engine when I can have a piece of software do it cheaply or even free in a matter of seconds? That's simple. I submit sites by hand because that's what the search engines require. Not "request" but *require*.

Using a piece of software to submit your URL to Bing and Yahoo! might save you a few minutes work, but it could cost you the ability to get listed. And let's be reasonable, they are ready to send you thousands of dollars in free sales and all they ask is that you abide by their requirements and hand-submit your URL. It's a no-brainer. Get typing or have someone do it for you.

SE⊙ 3.0 Submit Only As Needed	Rule #13

How your site becomes indexed can make a big difference in how Google later perceives it (and assigns value to it).

1. The best scenario is to have your site automatically spidered by Google after it has followed a good link in.
2. Second best is manually submitting your URLs.
3. Worst is using automated software.

Make SEO Copywriting a WOW!

The ability to craft text to not only satisfy the needs of both a search engine and live visitor is not enough in today's world. You have to be able to WOW them. There is far too much competition to merely interest the reader, be that reader a person or a bot.

Several years ago, Google introduced its natural text algorithm (NTA) and the game was forever changed. Before that, it was easy to write just what a search engine bot wanted to see and gain top rankings for your efforts. Not so today. Nowadays you need to walk a precarious line between what the search engine needs to see and what your visitor wants to see. And this is a good thing.

Do you remember some of those top-ranking pages from the early 2000's? What a bunch of junk! Here's an example of then versus now...

Then:

"Plumber Boston available for emergency services plumbing for plumbing problems from master plumbers Boston."

Now:

"Looking for an emergency services plumber in Boston? Look no further! Our certified and bonded master plumbers will have your plumbing related

problems fixed in no time. We are punctual, cour-
teous and always happy to be of help."

What a difference! Who would you rather call: the robot or the
nice friendly plumber? Google knows the answer and that's why
they added their natural text algorithm. They value the user - and
thus their experience - on any site shown in their results.

All this said, if you ever have to choose between two seemingly
equally qualified SEO people, choose the better writer. That bet
will pay off in spades every time.

SEO **Find Great Writers** *Rule #14*

The next time you are hiring, or re-evaluating current employee job duties, make it a priority to identify a good writer.
Then be prepared to eventually develop him or her into an
SEO copywriter. Again, I say *eventually*. This is not something
that will happen overnight but copywriting (of any kind) is a
skill set that pays for itself many times over - and SEO copywriting accounts for about 25% of your SEO success.

Not All Keywords Are Created Equal

A keyword is just another name for a search term. Basically it's anything your potential customer is typing into the search engine to find (hopefully) your website.

A GEO is simply a geographic point of reference, such as a town, city, or state.

When properly combined, keywords and GEOs are what targeted search engine traffic is all about.

For example, if you have a pizza parlor in New York City, you would want to be very specific about your business type (product or service) – in this case pizza, as well as your location.

The keyword will most likely be "pizza", unless you specialize in a particular type of pizza for which you want to be known. But the GEO might take some thought. Let's work through it…

"New York" is the state name but also a shortened version of the city name; aka "New York, New York." But is it more? In this case, yes. New York *style* pizza is very popular, sold all over the country and often simply called "New York Pizza."

Using one of several available tools, a good SEO professional would start by looking at the potential volume in terms of monthly traffic you could expect. Let's try it for the keyword "New York pizza" using Google's keyword tool.

Keyword	Jan. 2011 Search Volume
New York pizza	145,000

"New York pizza" search volume exported from Google's keyword tool

Using Google's keyword tool, we find that there were 145,000 searches performed in the previous month for that keyword.

Wow! That's a lot of potential visitors if you nail a top spot. But how targeted will it be? Let's look at the competition. A quick search of Google shows that we have 22,000,000 competing pages; many about New York *style* pizza, not about pizza actually located in New York.

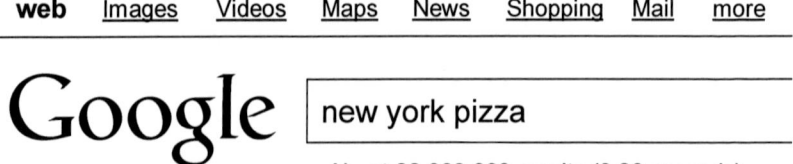

"New York pizza" Close-up as it appears in Google

A bit more research shows us that "New York City pizza" has monthly potential traffic of about 63,000 and much less competition at 6,480,000. Now, that's a big difference!

Keyword	Jan. 2011 Search Volume
New York City pizza	63,000

"New York City pizza" Search volume exported from Google's Keyword Tool

web   Images   Videos   Maps   News   Shopping   Mail   more

Google | new york city pizza

About 6,480,000 results (0.28 seconds)

"New York City pizza" Close-up

We still have plenty of potential visitors but now with only about one-third the competition (6,480,000 compared to 22,000,000).

Let's try again. How about "NYC pizza?"

Wow! 58,000 monthly searches and only 2,990,000 competing pages. Talk about a no-brainer!

Keyword	Jan. 2011 Search Volume
NYC pizza	58,000

"NYC pizza" Search volume exported from Google's Keyword Tool

web   Images   Videos   Maps   News   Shopping   Mail   more

Google | nyc pizza

About 2,990,000 results (0.28 seconds)

"NYC pizza" Close-up shows the lowest competition yet

You get the idea. Not all keywords are created equal. Consider these three possible scenarios...

- Some might bring you thousands of people who are not even looking for what you have to offer.
- Others might bring in all the business you can handle.
- Still others might attract a good number of visitors interested in a free offer but not becoming a paying customer.

Knowing how to spot the best keywords is one of the most important things your SEO company can do for you – or that you can do for yourself if need be.

There is so much more to keywords than we can cover here. The topic could easily fill an entire book itself – and actually has many times over. Just know that a true SEO expert has spent years sharpening his or her skills in this area - and for good reason; keyword selection is crucial to your success.

SE◎ Become a Keyword Curator — Rule #15

It can take just as long to optimize for a keyword that will bring in 10 visitors per month as 1,000. I'm not talking about "junk" keywords that some lazy SEO person tossed in the mix because they were easy to rank for. I mean REAL keywords with good search volume every month - and real competition. You need to become a keyword curator and get rid of those with low ROI so you can focus on those that bring in the profits.

A Note on Competition

The number of competing pages shown by Google is not necessarily a true indicator of actual "head to head" competition. It is however a true indicator of how many other pages have been (and remain) indexed by Google.

So what does that mean? It's simple, all of these pages might not be stiff competition, but they are all potential contenders to be recognized for that keyword - at least as far as the search engine is concerned.

Simply put, Google believes these page are relevant to the keyword searched. That in itself is competition, even if much of it is somewhat weak by comparison (if those "competing" Web pages have not been carefully optimized).

For example, do you suppose there are really twenty-two million Web pages on the Internet that are trying desperately to be number one for the keyword "New York pizza"? It's extremely doubtful.

However, of the 22,000,000 results (potential competing pages) shown on the following page for the keyword "New York pizza", even if only the tiniest percentage (say 1%) was actually optimized, that is still a great deal of tough competition (220,000)!

web   Images   Videos   Maps   News   Shopping   Mail   more

Google new york pizza

About 22,000,000 results (0.28 seconds)

Change Happens...Deal With It

Although it might be tough to get an occasional note that says "Heads-up – we've hit a snag." It's a whole lot better to get this than it is to get blindsided with a forty percent drop-off in sales.

If you ever receive an email like the following, embrace it as a chance to talk things over with your search engine marketing expert…

"We just experienced a major change in the search engines that has temporarily affected your rankings. Without getting into too much technical detail, Google has made a change to its algorithm (its formula to determine where your web pages rank for particular keywords) and you might notice a slight decrease in traffic while we make the changes necessary to counter any lost placement. As always, please feel free ask any questions – even the technical ones. Or if you prefer, we can set up a meeting for a time that is most convenient for you."

"Although these changes are temporary and we feel every confidence in being able to recover quickly, we wanted to make you aware of the situation so you can properly plan any mitigation strategies you might wish."

Now that's an honest person! Someone who you want to do business with. Not some techno-babble wielding idiot who treats you like a child and keeps you on a "need to know" basis – or just can't be bothered with you at all.

Just remember, change is inevitable and not always for the best at first. Anyone who tells you different is trying to sell you something, so run – screaming – away and don't look back. Do business with the one who dares to be sincere and you can minimize the nasty surprises that your competitors run into from their cheesy, self-proclaimed "experts" and "gurus."

In closing, embrace the honesty – even when it's hard to hear – and avoid techno-babblers like the plague. Even if your rankings temporarily suffer due to a mistake your SEM company made, it's better to deal with an honest human who makes mistakes than a weasel who tries to shift the blame or hide the issue altogether. Keep in mind that the occasional mistake is just as inevitable as change.

SEO(3.0) **Become an SEO Partner** *Rule #16*

Take every opportunity to partner with your SEO company or specialist. This is especially important when issues arise. You may never have one, but if problems do come up, remember that it only takes a moment to become part of the solution. And keep in mind that when you are fully onboard, you will be amazed at how much good can come from a seemingly bad situation. In SEO, "*challenges*" really are **opportunities**.

Get Social and Be Famous

Many SEO professionals have been saying for years that social media has no benefit to search engine optimization efforts… That's just crazy.

Any website considered an authority by Google is like gold. And considering that Facebook gets more daily visits than Google, I'd say that puts them in the 'golden' category. But it's not just Facebook; Myspace.com has been helping web pages rank higher for years. Twitter also pitches in, as do others.

But it's not just "active" social medial sites like these. Video sites like YouTube and Google Video offer a whole new level of what I classify as "passive" social media. Links from these (in the video tags) can prove extremely valuable your SEO efforts.

While there is no exact magic number of how greatly links from any type of social media site help, even .01% (.0001) makes it worthwhile. Yes, really…

Think about it... The average in-demand keyword today has about 1,500,000 competing pages vying for the top spot, depending on the niche and GEO.

That's 1,500,000 pages x .0001 = **150** competing web pages.

Simply put; where would you rather be in the search results, on page 1 or on page 15...150 spots away from number one?

You may not have the time or budget for this, or the next few tips, but I felt they should be included so you had the greatest potential advantages, whether you can immediately take advantage of them or not.

SE🅞 ³·⁰	Be More Sociable	Rule #17

As time and budget allow, get more and more involved in social media. Although you will want to research the best outlets at the time you do this, a good general plan includes (1) Setting up a Facebook fan page and getting as many people to "like" it as possible. (2) Gaining followers on Twitter to spread your word (and links) virally. (3) Earning inbound links from Facebook and Linked-In users - at a minimum.

Let the Blogging Begin

Please Note: The rule included in this chapter is supplemental and included only to take your SEO efforts over the top. Unlike the other rules, this may require additional time/money/effort.

Blogging is by no means a new idea, but it's amazing how many business owners and webmasters have ignored it, thinking it is something that only kids and bored people do. The reality is it can be one of the best marketing tools in your toolbox.

The concept is simple enough. You set up a blog (short for "weblog") using any free or paid method you like, such as WordPress. Then you can simply add content, on the fly, whenever (and from wherever) you wish. That content is made public for anyone to read and comment back on – unless you set restrictions.

So what's the big deal? The main benefits are its simplicity, convenience and power as an SEO tool. Let's break each down...

Simplicity: Once the blog is set up, all you have to do is type in a new entry whenever you get a few minutes and click "post." It's that simple. And search engines LOVE the fresh content.

Convenience: There is no big complicated administrative process to tie you up in knots. Just click a link to log in, enter your password (unless you have it saved in your browser) and start typing – from anywhere (note the word 'browser' earlier.)

Power as an SEO Tool: Blog posts are the ultimate in ethical spider-bait. When you have an active blog, search engine bots tend to come back and visit on a more regular basis. And did I mention that users can post replies if you so allow? Each one is even more content for the spiders to eat up. What could be better than getting free search engine optimization from the very people you are trying to reach?

SEO③.⓪	Get Blogging	Rule #18

Budget permitting, add a blog to your site and start reaping some great "social rewards" almost immediately. It's easy and costs no more than time if you use a free blogging engine like WordPress (wordpress.org). The SEO benefits will follow soon as long as you blog on topics related to your industry and be sure to feature important keywords prominently in your posts.

Go Mobile

 **Please Note:** The rule included in this chapter is supplemental and included only to take your SEO efforts over the top. Unlike the other rules, this may require additional time/money/effort.

I 'll keep this short and sweet. Google is heavily involved in the mobile market (think 'Droid). They know that mobile apps are fast becoming a mainstay of programmers, who are opting out of the PC rat-race for greener pastures. That said; they also know that more and more people rely on their mobile devices everyday for things they used to use a home computer for – including search.

There is a great deal of money to be made in mobile computing today, and mobile search tomorrow. Plan ahead and get in on a great thing at the ground level. It will not take much.

In reality, you just need a few mobile pages woven into your existing site. You can make things as easy or complex as you like, but I suggest easy – at least for starters.

The quickest and easiest way to get in on this game is to include some content pages in a mobile format (320 by 240 pixels is usually a good place to start but your developer will know for sure).

Be sure to exclude complex, and/or resource hog elements like Flash. And if you have images on any mobile page, include alt tags to save your visitor some frustration (and rank better).

That's about it. This one is simple. If you can afford to add a few mobile pages to get things going, that would be ideal. The sooner Google indexes them, the better off you will be.

SEO 3.0	Go Mobile	Rule #19

If your SEO and/or web development budgets allow, add a handful of "Mobile" pages to your site formatted for smart-phones, PDAs and other mobile devices (320 by 240 pixels is a common screen setting). Mobile compatibility will become increasingly important for Google and (even more) precedence is likely to be given to those sites that can accommodate mobile searchers. *(Think Android - Google is heavily vested.)*

Invent What You Cannot Find

 **Please Note:** The rule included in this chapter is supplemental and included only to take your SEO efforts over the top. Unlike the other rules, this may require additional time/money/effort.

If you need something that you cannot find, invent it – or have someone else invent it for you. This is a concept that goes beyond search engine optimization, but is still quite relevant here. Allow me to explain…

I get calls from clients all the time saying, "Is this possible?" or "Is there a way to do that?" Ninety-nine times out of a hundred, my response is "You bet it is!" – Or at the very least, "I'll find out." And these are questions that lead to developing innovations in search engine optimization.

It's only January of 2011 and I have already obtained two "patent pending" statuses – this year. Better still, I plan to have a dozen more by years' end – all on behalf of my company, my customers, or both.

Why is there a need to invent? Well, to be honest, there isn't really. Not usually, anyway. But at those times when the latest and greatest thing on the market won't do the trick, inventing is an excellent way to get exactly what you need.

Here's an example...

Problem: The need for a grand-scale website that can house hundreds of individual businesses, all sharing a single brand, and help each be found on the major search engines by keyword and GEO – WITHOUT customer crossover.

But wait. There's more... The site has to be easy to maintain for people with limited technical experience, yet at the same time, extremely appealing to customers.

One more thing... The users need to have select administrative rights to add pages and even GEO's at the push of a button; and I mean literally the <u>push of a button</u>.

And I almost forgot one more small detail. This site has to be self SEO-ing. This means that every time someone adds a new page, the site has to automatically optimize that page, and all of its sub-pages, for every major search engine. And not just any optimization. This has to be for both keyword and GEO depending on what the user wants.

Solution: *Hyper-Local* ™

In just a few short months we developed a solution called *Hyper-Local* ™ that worked so well, and whose technology was so innovative, that it is now patent pending.

SE○3.0   **Invent**   **Rule #20**

Last, but certainly not least, invent the thing you so desperately need but cannot find. Or have someone do it for you. Just get creative when the need arises. You will find that a great tool, program or process can give you a huge competitive edge in your market. As a bonus, you might just find innovation to be as addictive as it is rewarding.

Summary

This section is merely for the sake of convenience so you don't have to rummage through eighty pages to find the one rule you need at a moment's notice.

Feel free to photocopy these pages and keep them in a convenient location for easy access anytime you need the information.

SE☺ **Remove Your Black Hat** *Rule #1*

Before you do anything else, have a knowledgeable webmaster or SEO person go through your site and get rid of any black hat techniques that might be left over from the old days. This is critical to your SEO success.

Pre-Press Update: *As this book goes to press (January 25th), Google has just announced a new initiative to crack down on all SEO spam techniques.* **Check your site before it's too late.**

SE☺ **Welcome Change** *Rule #2*

Merely accepting change is no longer enough. If you want to succeed in the fast-paced world of search, you need to embrace it. Once you can anticipate change, based on experience and newfound knowledge, you are in a position to take the lead when opportunity arises.

Just remember that change can lead to self-fulfilling prophecy. If you expect problems, you will most certainly find them.

SE☺ (3.0) Evolve or Get Left Behind *Rule #3*

In SEO, or any online marketing, there is a cold, hard fact that must be faced... ***Those who do not evolve, get left behind.***

The web is not like print advertising or direct mail where yesterday's techniques will work for years - or even decades. Everything online happens FAST. You need to evolve *quickly* just to keep up and *willingly* - even <u>eagerly</u> - to get ahead.

SE☺ (3.0) Forget Clicks. Get Leads! *Rule #4*

Clicks are worthless unless they result in leads or sales. For most companies offering a product or service over $250, the goal is to get a qualified lead in order to close the sale - with the understanding that a qualified lead is the halfway point to a sale. That said; cut your sales legwork in half by focusing on leads over clicks. The "how to" of it could fill an entire book but simply asking your next SEO person "How can you get me leads instead of just clicks?" only takes a moment.

SE☺ (3.0) Make Bing a Priority *Rule #5*

Bing is no longer a *"nice to have"*. It is a **must have** unless you want to completely ignore one-third of your potential visitors.

Their deal with Yahoo! is big news but more importantly is the reason behind the deal. Microsoft (Bing) sees search as something they want to be in - in a very BIG way. How often do they lose when they really want something?

SE⊚ 3.0 — Get Rid of Your Junk — *Rule #6*

Junk keywords cost you time and money, so get rid of them. Now this does not mean highly targeted, yet low volume keywords with potential. This means absolute junk that nobody is searching for. So how can you tell? I subscribe to several services, but Google offers a nice freebie at:

https://adwords.google.com/select/KeywordToolExternal

SE⊚ 3.0 — Get in Google Places — *Rule #7*

Google Places is <u>going places</u> and your website needs to be along for the ride. If you have never looked into it, and claimed your free business directory listing, now is the time. It takes a phone call, possibly a postcard, and some patience, but is so worth it. As you are about the see; this is one of your most important "to do's". Visit at http://www.google.com/places/ or have your SEO person do it for you.

SE⊚ 3.0 — Place in Place Search — *Rule #8*

Times have changed. The old "Google Places" results are no longer hiding in the corner, hoping to get noticed. Now they DOMINATE the first search results page in a section called *Google Place Search*. Your potential visitors can't miss them so you cannot afford to miss out on being found within them. To do this you must first claim your Google Places Directory Listing in and try to build up some good reviews.

SE🔵 Take The 2 Second Test
Rule #9

Now that people no longer need to visit your website to see if they like it, you need to look at it in a whole new way - with fresh eyes, and for no longer than 2 seconds. That's how much time the average person will spend looking at it before they commit to the click. So print out some color screenshots of your home-page and a few of your top online competitors and honestly choose which is best - with just 2 seconds to decide.

SE🔵 Use Analytics
Rule #10

Having the information provided by a good website analytics program is crucial to long-term SEO planning and success. This can be a free app like Google Analytics (GA) or a paid app such as WebTrends. Just select one that provides, at a minimum, traffic stats, keyword tracking, goal tracking, and the ability to get custom reports at the push of a button. There are plenty on the market, with GA being the leader.

SE🔵 Set Goals & Track Results
Rule #11

Using Google Analytics, setting goals and tracking results is insanely easy - and free! Do it and reap immediate rewards.

Advanced Tip: If you have a multi-part form (such as a 2 or 3-step sign-up process) you can quickly add a funnel to learn just about everything you ever wanted to know about where conversions come from, and where they go to die.

SE**③.⓪**   Track Your Calls   *Rule #12*

If the business you receive via phone calls is at all important, invest in call tracking. It's worth the extra $40 or so per month. In fact, *it's too cheap not to use*. When properly set up it's like having a 24/7 operator showing you exactly how each piece of new business came in. It will even let you pick the search engines you want to track calls from, whether the calls were generated by organic search or ads, etc. This is a BIG deal.

SE**③.⓪**   Submit Only As Needed   *Rule #13*

How your site becomes indexed can make a big difference in how Google later perceives it (and assigns value to it).

1. The best scenario is to have your site automatically spidered by Google after it has followed a good link in.
2. Second best is manually submitting your URLs.
3. Worst is using automated software.

SE**③.⓪**   Find Great Writers   *Rule #14*

The next time you are hiring, or re-evaluating current employee job duties, make it a priority to identify a good write. Then be prepared to eventually develop him or her into an SEO copywriter. Again, I say *eventually*. This is not something that will happen overnight but copywriting (of any kind) is a skill set that pays for itself many times over - and SEO copywriting accounts for about 25% of your SEO success.

SE⊚ Become a Keyword Curator — Rule #15

It can take just as long to optimize for a keyword that will bring in 10 visitors per month as 1,000. I'm not talking about "junk" keywords that some lazy SEO person tossed in the mix because they were easy to rank for. I mean REAL keywords with good search volume every month - and real competition. You need to become a keyword curator and get rid of those with low ROI so you can focus on those that bring in the profits.

SE⊚ Become an SEO Partner — Rule #16

Take every opportunity to partner with your SEO company or specialist. This is especially important when issues arise. You may never have one, but if problems do come up, remember that it only takes a moment to become part of the solution. And keep in mind that when you are fully onboard, you will be amazed at how much good can come from a seemingly bad situation. In SEO, "*challenges*" really are **opportunities**.

SE⊚ Be More Sociable — Rule #17

As time and budget allow, get more and more involved in social media. Although you will want to research the best outlets at the time you do this, a good general plan includes (1) Setting up a Facebook fan page and getting as many people to "like" it as possible. (2) Gaining followers on Twitter to spread your word (and links) virally. (3) Earning inbound links from Facebook and Linked-In users - at a minimum.

SE🌐   Get Blogging   *Rule #18*

Budget permitting, add a blog to your site and start reaping some great "social rewards" almost immediately. It's easy and costs no more than time if you use a free blogging engine like WordPress (wordpress.org). The SEO benefits will follow soon as long as you blog on topics related to your industry and be sure to feature important keywords prominently in your posts.

SE🌐   Go Mobile   *Rule #19*

If your SEO and/or web development budgets allow, add a handful of "Mobile" pages to your site formatted for smart-phones, PDAs and other mobile devices (320 by 240 pixels is a common screen setting). Mobile compatibility will become increasingly important for Google and (even more) precedence is likely to be given to those sites that can accommodate mobile searchers. *(Think Android - Google is heavily vested.)*

SE🌐   Invent   *Rule #20*

Last, but certainly not least, invent the thing you so desperately need but cannot find. Or have someone do it for you. Just get creative when the need arises. You will find that a great tool, program or process can give you a huge competitive edge in your market. As a bonus, you might just find innovation to be as addictive as it is rewarding.

Before You Invest a Dime

I have talked a good deal about things to do and things to avoid, but I have not discussed *how to get the best deal on your search engine optimization work.* Here goes...

First of all, <u>know what to expect</u>. SEO is not cheap, but it can be a tremendous bargain. You may spend a few hundred to a few thousand dollars per month, depending on your needs, and that is quite reasonable – as long as you get more out of it than what you put in to it. It's all about return on investment, or ROI. I'll show you what I mean...

One of our small business clients invests $595 per month for the SEO of his home improvements website. It was a big leap of faith to try SEO (reassigning 75% of his marketing budget) but it paid off. *At last report he was getting three times the number of leads from SEO as all other marketing methods combined – for less money.*

On the far side of the spectrum, one of our larger clients invests several thousand per month and considers it "pennies on the dollar" based on return – which in their case is sign-ups.

Their traditional and PPC advertising had cost an average of $120 per sign-up throughout 2010. Their SEO cost per sign-up? $7. No, that is not a typo. Seven dollars compared to one hundred twenty dollars! And the best part; SEO delivered eight times (8x) the sign-ups at about *6 cents* **on the dollar!**

Did they stop using the other methods? Not at all. They want all the sign-ups they can get. At a cost of $120 per sign-up they make a small profit but at the SEO cost of just $7 per sign-up they make a huge margin, which allows them to do even more with other mediums. It's a win, win.

Next up; <u>find out what *exactly* is being offered</u>. Ask for sample keywords and then ask *why* these are the best ones. In fact you should <u>get a sample of their work</u> to make sure they know what they are doing. ***Just ask for a free evaluation of your site.*** No legitimate SEO company should refuse.

At NetSearch Direct we do a complete *"Search Engine Snapshot Report"* on every new potential project, which includes:

- Target Market Data
- Keyword Research
- Current Ranking Report
- Estimate of Leads you could expect from SEO/PPC
- Local Business Directory Listings Analysis (if applicable)

All are extremely valuable for any online marketing efforts. If you would like a ***free Search Engine Snapshot Report***, just contact me at 804.228.4400 or MikeM@NetSearchDirect.com or fax me the form at the back of the book.

Frequently Asked Questions

For this section, I wanted to add some info on search engine marketing (SEM) as well as SEO. As you are the one plotting your company's patch to online success, I figured you might like to know what all of your online marketers are talking about. And, since my company, NetSearch direct, also specializes in SEM and Social Media, we have accumulated hundreds of questions on just about every topic imaginable.

To show every relevant question NetSearch Direct has ever been asked all would take a much larger book. That said; I included only those that were asked multiple times, by multiple people, thus making them true "frequently asked questions" and not just random bits of knowledge.

Even still, there is a LOT so I've made each FAQ entry as concise as possible to give you quick and accurate answers without any fluff. To make it even easier, you will find each FAQ topic listed alphabetically. You won't even have to read "What is" two hundred times. Just look up the actual item you are interested in, like a glossary but with extra info to put it in context.

Here we go. And, just to set the tone:

What is..?

AdSense

Google's AdSense program is a free-to-join advertising revenue-share program that allows publishers (typically website owners) to earn money by displaying AdWords ads on a wide variety of online content resources including site search results, websites, feeds, and mobile web pages.

AdWords

Google's AdWords is a "pay for placement" text advertising program. You can display your ads on Google and you only pay if people click on them (cost per click) – or by the number of impressions shown (CPM) – as you choose.

You can create ads and use keywords based on your business to attract more customers through searches on Google or any relevant resource you like in the AdSense network. It's simple. People search for an item or service by using a keyword. If you bid on that keyword (and bid high enough) your ad is shown and you either pay per click or by the number of impressions. Although PPC is the most common method advertisers choose, Google has been known to experiment with alternative methods such as "cost per action" (CPA) that has the advertiser paying only when a call to action has been completed.

Algorithm

An algorithm is like a recipe. You follow steps in a particular order, using only the correct ingredients, to get the desired results. For our purposes however, algorithms are used to determine how search engines index content and display the search results to users.

Alt Tags

Alt tags are an alternate text tag that is inserted into a web page graphic (picture) so that text is displayed when an Internet user hovers the mouse over the graphic. The idea behind alt tags was originally to provide a descriptive alternative to viewing image files for those users who had their images turned off in their web browsers. This was more common that you might think back in the days of dial-up Internet access when a typical web page load time was often counted in minutes rather than seconds.

Alt tags should contain relevant keywords and describe the graphic. The value of Alt tags for SEO has been discounted over time by the search engines to the point that now it is of minimal importance.

Anchor Text

Anchor text is the clickable text in a hyperlink. Years ago, it was easily detectable as nearly always being blue and underlined. That is no longer the case. Now, it can be any color or style and even appear as regular text your mouse has to trip over before you realize it is a link (when your mouse hovers over anchor text, the destination URL appears).

The actual text used is important because search engines use it as an important ranking factor. Google pays special attention to the text used in a hyperlink and associates the keywords contained in the anchor text to the page being linked to. It's a big part of the popularity contest method of ranking used by Google.

Automated Submitting or Submission

Automated submission is just like it sounds; it is automatic software or an automated service to submit your web pages to the search engines.

In the 1990's it was no big deal because search engines were hungry for more pages to fill their SERP's (search engine results pages). But, today this tactic is frowned upon by all major search engines and can land you in hot water if you try it.

Backlinks

Backlinks are incoming links pointing to a web page. The number of and Google PageRank (PR) of backlinks is an indication of the popularity or importance of the web page being linked to. They are also known as incoming links, inbound links, inlinks, and inward links.

Bait and Switch

Bait and switch is a technique where the developer creates an optimized page and submits it to search engines. Then, once the optimized page has been indexed, it gets replaced by the much more attractive user-friendly page.

Bid Management Tool

A bid management tool can be either software or a service used to manage bids on pay-per-click search engine advertising programs like Google AdWords. If you have been in the SEM game for a while you might recognize names like BidMax, BidRank, and KeywordMax as being some of the more popular.

Bidding (or Keyword Bidding)

No big surprise here. Bidding means placing a bid price that you are willing to pay as an advertiser on a pay-per-click search engine. The highest bid for a given keyword achieves the top spot in the PPC search results – with some exceptions.

Google throws "a monkey in the wrench" with its whole ad quality score deal where you bid to be in a group (say, the top three places) and a complex method to determine your actual placement is used, such as ad relevance to the target URL page, wording of the ad, and popularity of the ad based on click through rate.

In other words, you might pay more to be number three than the person who is showing up as number one. It all depends on your quality score.

Black Hat SEO

The term "black hat" is taken from old black and white westerns where the good guy was easily identifiable by his bright white hat and "Black Bart" was always dressed like Johnny Cash with the addition of a matching hat. So now the name makes more sense...

Black Hat SEO is any optimization tactic that causes a site to rank more highly than its content would otherwise justify – until it gets busted.

As its name indicates, these techniques are considered unwelcome by the major search engines. However, they are still used today, more than ever, due to the lure of easy rankings and ease of starting over if busted. Be warned however, it's not that easy. Some search engines, like Google, will make it their mission to make sure offenders find it extremely difficult to ever rank again in their network, and if you make the blacklist – any search network.

Blacklist

A blacklist is literally a list of those websites banned by search engines, for whatever reason. Most major search engines compile their own lists, consisting mostly of spammers and those who use other black hat tricks. But, there's more… These lists get shared across search networks and it's not just a list of websites anymore. Now the list can include IP addresses and even personal information about the site owner.

Blog

The term "blog" is a relatively recent addition to online vocabulary and is just a shortened version of "web log" because web log is so much more difficult to say (those darn lazy kids today...).

These started out as basically just online diaries with entries made on a regular, if not daily, basis. But, today they are so much more. A blog is a soapbox from which anyone with something to say can be heard. A couple short decades ago, getting the public's attention was reserved for movie stars, politicians, and a few people who had something to say that captured the media's attention and promised strong ratings. Nowadays, a blogger can get their own personal page for free and start reaching out to the world in mere moments.

Blogs are also great for SEO purposes due to their fresh content and (typically) simple navigation structure.

Body Copy

The body copy is the main text or content of a web page. More precisely, it is the text visible to users and does not include graphical content, navigation, or information unseen in the HTML source code.

From an SEO perspective, body copy is very important. This is a major part of what search engines use to decide how relevant your web page is when it comes to determining its placement in the search results.

Bot

Bot is short for robot. Also known as "spiders" and "crawlers", they are programs used by a search engine to explore the web by following all available links. They do not work alone, however. Once a bot has found a page to scan it works with an indexer to download the HTML content and store it in a database that returns the search listings.

Broad Match

Broad Match is a form of "keyword matching" used in pay-per-click advertising campaigns. It refers to the matching of a search listing or advertisement to selected keywords in any order. Broad match terms are less targeted than "exact" or "phrase" matches.

For example, if a person searches for "Michelin brand tires" and your ad campaign is set to "broad match", your ad is eligible to be shown for "Michelin" or "tires" or "brand tires," etc. – in addition to the full search term of "Michelin brand tires."

Cache

A cache is simply a copy of a web page stored locally on an Internet user's hard drive or within a search engine's database. It is the reason web pages load so quickly when a user revisits a page... It loads instantly because the page is not being re-downloaded off of the web - it is already stored in cache.

From an SEO perspective, it's also a handy way to see what version of the page has been indexed on a search engine like Google. By simply clicking the "Cached" link just below the listing you can see the last version of the page Google indexed.

Call to Action

A call to action is any method used in advertising to encourage a person to complete a task as defined by the advertiser. It is usually a combination of text and graphics that encourage the user to complete a form or click a link. You see them all the time with words like "Click here," "Buy now," "Enter now," or "Click to download."

Clickthrough

A clickthrough is the action of clicking on a link and causing a redirect to another web page. It is basically the act of clicking "through" a link to get to the "offer" or "end result" page.

Clickthrough Rate (CTR)

As with all things associated with advertising, if it's not measurable, it's not meaningful. That said, an ad's clickthrough rate is basically its success rate. This is determined by a simple formula of dividing the number of clicks to an ad by the number of times that ad was served (seen by users). ***Note:*** *The number of times it is served is known as the number of impressions.*

CTR Formula: CTR = Clicks / Impressions

So if there were 1,000 impressions served, and that ad received 10 clicks, the clickthrough rate would be 1% (10 clicks / 1,000 impressions = .01 CTR).

Cloaking

Cloaking is the act of showing different content to search engine spiders than to human visitors. This black hat technique sets up "follow" links for specific search engines and "no-follow" links for the pages they want only humans to view (often sneaking in redirects from optimized pages to really take advantage).

As you can well imagine, search engines frown upon this practice. While some will merely penalize an offending website, others ban them completely.

Conversion

A conversion in online terms is basically the same as in traditional offline terms. The act of transforming a website visitor into a customer is "customer conversion", while the act of taking that visitor a step closer to becoming a paying customer can be considered a "lead conversion".

Conversion Rate (CR)

Conversion rate, much like clickthrough rate, is the rate at which visitors get converted to customers or leads (please see above).

CR Formula: CR = Goals Completed / Number of Visits

So if there were 10 goals completed, after 1,000 visits, the conversion rate would be 1%

$$10 \text{ goals} / 1,000 \text{ visits} = .01 \text{ CR}$$

Cookie

A cookie is a tiny bit of information, in the form of a text file, which is placed on a visitor's computer by a web server. While the website is being accessed, data in the visitor's cookie file can be stored or retrieved.

Some of these, called "session cookies", will disappear once you close out your browser, while others will stay in your hard drive to be ready for the next time you visit that page. They are simply there to identify you to the web page you are visiting. This is meant as a convenience to eliminate the need for you to re-select user preferences, etc.

Cost Per Action (CPA)

Cost per action is the expense incurred or price paid for a specific action, such as signing up for an email newsletter. If you pay $1 per click and it takes 20 clicks to get a visitor to complete an action, then the CPA is $20.

Cost Per Click (CPC)

This is the price paid for a clickthrough of your ad in PPC.

Cost Per Lead (CPL)

Cost per lead is simply the average price paid to attain a new lead. Its formula is very simple. Just take the amount spent and divide it by the number of leads generated.

If you get 20 leads from a campaign that costs $100, your cost per lead is $5.

Cost Per Order (CPO)

Cost per order is just like CPL; just replace "lead" with "order."

If you get 50 orders from an advertising campaign that costs $500, your CPO is $10.

Note: CPO is also known as cost-per-transaction, or CPT.

Cost Per Sale (CPS)

CPS is a representation of the sales revenue divided by total ad spend, with that number divided by the number of units sold.

Cost Per Thousand (CPM)

CPM is the price paid for one thousand impressions of an advertising campaign. It is a bit confusing because the "M" in CPM is a holdover from earlier marketing days when the "M" was used as a sort of shorthand for the Roman numeral "M", which is the number one thousand.

Crawler

A crawler is just another name for a "spider" or "bot", which are programs used by a search engine to explore the web by following all available links. They do not work alone, however. Once a crawler has found a page to scan it works with an indexer to download the HTML content and store it in a database that returns the search listings.

CSS (Cascading Style Sheets)

CSS is a "style sheet" language used to control the look and formatting of text elements of a web page. It allows a web designer to control what fonts are used, as well as their sizes and colors, for any declared value on the page. These are typically used for "heading tags" and paragraph text.

Using CSS is very efficient as it allows the designer to change multiple aspects of an entire website with a single update to a text file. It is a nice change from "hardcoded" manual style changes.

This is also handy from an SEO perspective as it allows the optimizer to manage the look of heading tags and other elements, so the user experience is as desired while the search engine spider experience is also optimized.

CTR (Clickthrough Rate)

An ad's CTR, or clickthrough rate, is its measurable success rate.

It is determined by a simple formula of dividing the number of clicks to an ad by the number of times that ad was served (seen by users) – also called its number of impressions.

CTR Formula: CTR = Clicks / Impressions

So if there were 1,000 impressions served, and that ad received 10 clicks, the clickthrough rate would be 1% (10 clicks / 1,000 impressions = .01 CTR).

Custom Error Page

You can customize the content including the look and feel of the default page that is displayed on your web server when a "404 File Not Found" error occurs. A good 404 error page has a friendly message explaining that the page they requested doesn't exist at the location, a site map to encourage the user to continue exploring the site, and a search box so the user can conduct a search.

Database-driven

Database-driven refers to any web page that extracts full or partial content from a database in order to populate it with (hopefully) relevant content.

This method has several SEO advantages. Chief among these is the ability to load (in proper proportion) any keywords and GEO's you would like to appear, on any relevant page, as part of an SEO baseline.

Database-generated

Database-generated refers to a web page that is actually created from the content within a connected database. A "dynamic" web page is a good example of one that is database-generated.

Directory

A directory is similar to a search engine in that it provides results for items searched on the web. It differs however in that (in a directory) human editors group websites into categories and provide site descriptions or edit descriptions that are submitted to them. In a search engine, this is all handled by bots and indexing software that run constantly.

Yahoo! Was the original directory and continued to post directory results until the mid 2000's.

Doorway Page

A doorway page is typically a low quality web page that contains very little content and exists solely for the purpose of driving traffic to another page, although there are exceptions to the rule. In fact, some doorway pages actually contain more relevant data than the main page being directed to.

Dynamic

Dynamic pages are generated, as needed and depending on the user's request, from a database. In regards to SEO, these pages are especially helpful in capturing very specific targeted keyword driven traffic, including "long-tail" keyword searches (the main keyword plus some additional words making it more specific).

Exact Match

Exact match is a form of keyword matching where the search query must be exactly the same as the advertisement keyword. Let's use our Michelin example again…

If a person searches for "Michelin brand tires" and your ad campaign is set to "exact match", your ad is NOT eligible to be shown for just the terms "Michelin," or "tires," or even "brand tires." It is ONLY eligible to be shown for the full and exact search term of "Michelin brand tires."

This gives the advertiser an opportunity to be extremely precise in deciding what he or she is willing to pay for.

Exact Match Domain

An exact match domain is when the domain name exactly matches the keyword searched in a search engine or directory. This can be very powerful in terms of SEO advantage if the SEO person knows what they are doing.

Findability

Findability is a term referring to the measure of how easy it is to find a particular web page using search engines.

Firefox

Firefox is a web browser made by the Mozilla Corporation. Although Firefox is one of the newer kids in the neighborhood, Mozilla is a name recognized as one of the originals in the web browser arena.

Firefox is free and open source software developed by Mozilla and a community of external contributors.

Flash

Flash is a technology, invented by Macromedia, that allows a developer or web designer to embed interactive multimedia into web pages. This might be anything as simple as a short video to a multi-part interactive sign-up form.

Of course, one of the things that made Flash famous was its flexibility in making online games that are quick and easy to play (desktop toys).

It is typically best to avoid Flash from an SEO point of view as it adds no value to a search engine and can actually minimize rankings as search engines are not able to read a movie (or text within an image), just a few tags on the back-end if they exist.

Flash Intro

A Flash intro is an animated introduction, typically in the form of a video splash page (created using Flash of course) that Internet users often have to sit through upon entry to a website. This can be very annoying – especially to return visitors - so it is typically best to have a "skip intro" option.

Frames

The use of frames is a method of combining multiple web pages into a single "parent page". Each individual page is grouped into one, and potentially has its own scrollbar.

Note: You know you're on a framed website when part of the page scrolls while the rest of the page stays in place.

Frequency

Frequency, in terms of search engine marketing, refers to the number of times an ad is delivered to the same browser in a single user session (visit).

Fresh

Fresh is a common term adopted and often used by Google that describes regularly updated on-page content. The advantage to fresh content is that the spider will typically come and re-index the site more often.

GAP - Google Advertising Professional

Google Advertising Professional is an online marketing industry rating.

It is obtained via a free program, for professionals wishing to manage multiple Google AdWords clients.

Gateway Page

The term "gateway page" is often used as another term for a doorway page, which as we know, is typically a flimsy page created as spider bait, for the express purpose of driving traffic to the real home page. A gateway page can be more however…

In the mid 2000's a gateway page typically referred to a higher quality page with well-written, and unique, SEO content whose purpose was to act as a legitimate alternate point of entry meant to direct visitors to deeper, more relevant, parts of the site, rather than the home page.

Geo-Targeting

Geo-targeting is a method of very targeted advertising in which ads are distributed based on geographic location. Online advertising, like Google AdWords, allows for the geo-targeting of countries, states, cities, and suburbs.

You can also geo-target in SEO. This is when you optimize for a keyword that includes the geographic qualifier (IE/ city name) as part of the keyword. For example:

"New York City pizza delivery."

Google

Google is currently the world's number one search engine with the lion's share of the word's search engine market share (typically between 50% and 60%). Nipping at its heels, constantly striving to get a bigger piece of the pie, are Yahoo! and Bing.

Fun Fact: Founders Larry Page and Sergey Brin named the search engine they built "Google" as a play on the word "googol" the mathematical term for a 1 followed by 100 zeros.

Just for fun, here is a googol:

10,000,000,000,000,000,000,000,000,000,000,000,000,000,0
00,000,000,000,000,000,000,000,000,000,000,000,000,000,000

The name reflects the immense volume of information that exists, and the scope of Google's mission: "to organize the world's information and make it universally accessible and useful."

Google AdSense

AdSense is a free advertising revenue share program that lets online publishers to earn money by displaying relevant ads on their websites (and other online media). Here's how it works:

You, as the AdSense affiliate, place an ad block on your site. The ad block shows paid text ads that are relevant to your visitor by being relevant to your page content.

When someone clicks on the ads you placed on your site, you earn revenue. This is typically a cost-per-click revenue model, wherein you get paid x amount for every click. – although you usually never know the exact value of x until after you earn the money.

Google Analytics

This is a free web analytics tool, requiring nothing more than a small bit of JavaScript code on your pages, that offers a great deal of detailed visitor statistics.

Google Analytics can be used to track all the usual site activities: visits, page views, pages viewed per visit, bounce rate, average time on site, and much more.

Google Dance

The term "Google Dance" refers to those occasional times when Google indexes are updated. This almost always results in fluctuations in rankings and the displayed the index size, but usually settles back down within a short time.

Google Juice

Google juice is insider slang referring to the imaginary substance that flows between web pages through their links. It's basically the particular value that Google credits your page for having an inbound link. The value adds up for each link, and more juice means better rankings.

Google Supplemental Index

This is a database filled with supplemental results pages considered less important – or less trusted - by Google's algorithm.

Google Toolbar

The Google Toolbar is a powerful, and free, browser add-on. It is available for both Internet Explorer and Mozilla Firefox and displays info like PageRank and services such as news or email.

Google Traffic Estimator

Google Traffic Estimator is a tool that indicates the number of clicks to expect on Google AdWords ads for particular keywords. It is used to help the advertiser estimate search volume, average CPC, likely ad position, estimated clicks per day and cost per day.

Google Trends

Google Trends is a tool from Google Labs. It allows you to see how Google search volumes for a particular keyword have changed over a period of time. It shows the popularity of search terms and what people are searching for on the Internet.

Google XML Sitemap

Google Sitemaps are XML files that list the URLs available on a site. They help site owners notify search engines about the URLs on a website that are available for indexing (always a good idea).

Googlebot

Googlebot is a spider or search bot that scans and catalogs documents from the web to build a searchable index for Google to display in its search results.

Heading Tag

A heading (or H tag) is an HTML tag used to denote a page or section heading on a web page (kind of like a section title). Search engines pay special attention to text that is marked with a header tag – especially the H1 tag. These tags are also commonly called "H tags" and range from "H1" to "H6", with 1 being considered high priority.

Hidden Keywords

Hidden keywords are those placed in the HTML source so that they are not easily viewable by human visitors looking at the web page - but are viewable by search engines in the source code.

This is just another version of "hidden text" and usually means the words are simply the same color as the page background.

Hidden Text - SEO Spam Tactic

Hidden text is a SEO spam tactic to (visually) hide html text from human visitors in a web page. However, it is available for search engines to spider because it does exist – it is just the same color as the page background so people are not able to discern it.

Hits

A hit basically a "incremental tag" that counts every time a file is downloaded from a web server. Contrary to popular belief, most hits do not correlate directly with web page visits. Every image file that loads in a browser is a hit.

HTML

HTML stands for hypertext markup language and is the main coding language used in creating web pages.

HTML Source

This is the raw programming code. Visitors don't see it but search engines do..

HTTP - Hypertext Transfer Protocol

HTTP is an acronym for Hypertext Transfer Protocol, which is a networking protocol and the most basic component of data communication for the web.

You see it all the time but may not realize it *(http://domain...)*

Hyperlinks

Hyperlinks (aka "links") are an element (like text or graphics) of a web page, email message, etc., that when clicked, will redirect the user to another web page or other location (such as to a download file, etc).

Impression

An impression is one "serving" of a search advertisement to one user session.

Inbound Links (IBL)

Inbound links are those that point to your site from an outside source... In other words, they are links coming <u>to</u> your website <u>from</u> another.

They are an extremely important asset that have nearly unlimited potential to improve your site's rankings and PageRank.

Index

An index is a search engine's database where it stores the data (saved as text) from each page its spider visits. Once "indexed" the data is sorted and organized for future utilization as a search return.

Internal Link

An internal link is one that points to another page within the same site (as opposed to an inbound or outbound link). Their main purpose is to act as a navigation resource for people, providing quick and easy access to pages within the website.

Of course, search engines use these too, so having clear, easy to follow links is very important for SEO.

Internet Explorer

Internet Explorer is a web browser produced by Microsoft and is (no surprise here) the most widely used in the world.

Tip: It comes as part of the default Windows OS install so it's a bad idea to remove it.

IP Address

An IP Address (Internet Protocol) is basically the virtual identifier of any network device. It is broken into a four-part series of numbers separated by 'dots'. Each IP address identifies every sender and receiver of network data.

Think of this as being like a home address. Instead of State, City, Street, Dwelling Number, an IP addresses numbers represent the Domain, Network, Subnet, and Host Computer.

Every device connected to the Internet is assigned a unique IP. It may be permanent (aka static) or temporary (aka dynamic).

Note: many companies try to get a static IP address for their website as it has some SEO benefits – less now so than before.

ISP - Internet Service Provider

ISP is an abbreviation for *Internet service provider*. Most ISP's provides a wide range of Internet-related services to customers – not just Internet service.

These often include email, website hosting, domain name registration, and even merchant accounts. Think AOL circa 1998 if you want a good example.

Java Applet

A Java applet is a small program written in the Java programming language that can be embedded into web pages. Applets run on the user's computer rather than the web server's computer.

JavaScript

JavaScript is a scripting language that is a "lite" version of the complete Java programming language. A great example of JavaScript in use is the code snippet provided by Google for their Analytics tracking.

Keyword Phrase

A keyword phrase is a search phrase made up of any number of keywords. Most people use "keyword" and "keyword phrase" interchangeably because it is not often today that people optimize for a single "one-word" keyword.

Keyword

A keyword (officially called a query in tech terms) is a word (or group of words) that a search engine user uses to find relevant web pages.

Keyword Density

Keyword density is the number of occurrences that a given keyword appears on a web page. The more times that a given word appears on your page the more weight that word is assigned by the search engine.

Keyword Efficiency

Keyword efficiency is a measure of how effective a keyword will be when used in an SEO campaign.

This is such a big deal that there are a number of specialized formulas out there to help determine that efficiency, such as…

Keyword Efficiency Index

The Keyword efficiency index is a factor generated by using a KEI formula that takes into account the search volume and the number of competing web pages. It's used to measure the *potential* SEO impact of individual keywords.

But don't read too deeply into that. It is not a magic bullet to tell you what keywords will be easiest to rank for. Think of it more as "bang for your buck" indicator with many variations. The simplest and most common is as follows:

KEI $= (V)^2 / (P)$
V= Volume; the number of Monthly Keyword Searches
P= Number of Competing Pages

What this really provides is a quick indicator of if a keyword is worth going after. The higher the KEI factor, the higher the (potential) value.

Keyword Efficiency Index with Relevancy

Keyword efficiency index with relevancy (KEI R) is the same as the keyword efficiency index formula, only it includes the relevance factor of your keyword that you manually insert, to help determine its most accurate value.

Keyword Matching

Keyword matching is the process of selecting and providing advertising or information that match the user's search query. There are four types of keyword matching: broad match, exact match, phrase match, and negative keyword.

Keyword Popularity

Just like it sounds, keyword popularity is a term used to describe how in-demand any given keyword is, at any given time – based on how many people are searching for it (search volume)..

Keyword Prominence

This indicated the location of a given keyword on the web page. The higher up in the page (or closer to the beginning of the spiderable portion of the page) a particular word is; the more prominence it has. The more prominent the keyword, the more value that particular term is assigned by the search engine when it matches a keyword search.

Keyword Research

Keyword research is the due-diligence performed to determine the words and phrases that people use to find something online. Once determined, the (now) keywords are compiled into a list for use in the SEO efforts.

Keyword Stuffing

Keyword stuffing is placing an excessive number of keywords into a web page. It typically refers to having too many instances of one or more keywords rather than too many separate keywords. Here's what I mean…

Keyword 1, keyword 1, keyword 1, keyword 1, keyword 1, keyword 1, keyword 1, keyword 2, keyword 2, keyword 2

Not so much:

keyword 1, keyword 2, keyword 3, keyword 4, keyword 5, keyword 6, keyword 7, keyword 8, keyword 9, keyword 10

Keyword-rich

Keyword-rich refers to when a web page or chunk of text is full of good keywords rather than a bunch of meaningless words or irrelevant words.

Taking it a step further, it also means that each keyword is used an appropriate number of times compared to the amount of overall content.

Note: This is important!

Landing Page

A landing page is a web page where people go to once they click on an online advertisement or natural search listing. Landing pages are carefully designed to be extremely relevant to the advertisement or search listing.

Of course, it's easy to build a highly relevant landing page because you can design it to target just a single (or multiple) specific keyword(s) or theme(s). Just don't get this confused with a doorway page. It is much different.

Link Bait

Link bait refers to useful or entertaining web content that makes other webmasters want to link to it. Games and interesting stories are among the most popular.

Link Building

Link building is actively researching, finding, and obtaining links from websites for the purpose of increasing your "link popularity" and/or Google "PageRank" – and ultimately, ranking.

Link Farm

A link farm is a group of interlinked sites, usually managed by a single person or company that coordinates the swapping of links between members. The idea is to appear more popular to search engines. As you probably guessed, this is a big no-no

Link Popularity

Link popularity is how popular your link is or how many times other web sites have linked to your site. The more web pages that link to you, the better your link popularity will be, and thus rankings – at least that's the plan.

Link Spam

Link spam refers to irrelevant links between pages that are set up just to take advantage of a link-based ranking advantage offered by most search engines. Don't bother, it does not usually help and can actually hurt your rankings.

Links

See hyperlinks.

Log File

A log file is a record of information stored on the web server that typically includes date and time, files accessed, user's IP, web page visited from (referrer), browser type/version, etc.

Manual Submitting or Submission

Manual submitting, or submission, is when you submit your website by hand, for inclusion into an individual search engine, rather than using an automated submission tool or service.

Meta Description

Meta "description" is a tag in the HTML that describes the page's content. This is not visible on the actual web page but is very visible as the description in the search results – so make it good.

Meta Keywords

Meta "keywords" is a tag in the HTML that lists keywords relevant to the page's content.

Tip: Use only keywords that are visible on the page, otherwise it looks like spamming to the search engine.

Meta Search

Meta search is a group of search results from several sources that have been consolidated into a single SERP. Back in the late 1990's up until about 2002, these were very popular and search engines (using the term loosely) like Dogpile.com got a lot of attention.

Note: Meta Search sites generally do not have their own index or search database. They merely compile and present others'.

Meta Tag Stuffing

Meta tag stuffing is the act of repeating keywords in the Meta tags (like the *keywords* tag or *description* tag) and/or overusing Meta keywords that are not related to the site's content.

See the "Tip" under **Meta Search** for more detail on the latter as this has become a more important issue recently.

Meta Tags

A Meta tag is information that is associated with a web page and placed in the HTML but not visible on the displayed web page.

Mouseover

A mouseover occurs when a user 'hovers' the mouse over a link without clicking - and it displays something new on the page, such as a "text tip" or menu option previously hidden.

Mozilla Firefox

Mozilla Firefox is a web browser. Firefox is free and open source software developed by the Mozilla Corporation and a community of external contributors.

MSN

MSN is an acronym which refers to the Microsoft Network and their search engine, which changed its name to Bing a couple years back.

Negative Keyword

A negative keyword is a term from Google AdWords and is a form of keyword matching. It means that an advertiser can specify search terms that they do not want their ad to be associated with.

For example, if you add the negative keyword "-purple" to the keyword "flowers", the ad will not be displayed if a person searches upon the term "purple flowers."

Negative keyword matching seeks to ensure that only qualified traffic is eating up your budget.

Outbound Links

Outbound links are hyperlinks that point to another website (on a different host or IP) from your website.

Page Title

A page title simply what you call the page, inserted into the Title tag. It is also one of the more urgent parts of onsite optimization to get right.

Pagejacking

Pagejacking is the act of stealing high-ranking page content from another site and placing it on your website in hopes of increasing your own search engine rankings.

PageRank (PR)

There's no way I can improve on this so I'll let Wikipedia explain this one…

"PageRank is a link analysis algorithm, named after Larry Page, used by the Google Internet search engine that assigns a numerical weighting to each element of a hyperlinked set of documents, such as the World Wide Web, with the purpose of "measuring" its relative importance within the set. The algorithm may be applied to any collection of entities with reciprocal quotations and references. The numerical weight that it assigns to any given element E is referred to as the PageRank of E and denoted by PR(E).

The name "PageRank" is a trademark of Google, and the PageRank process has been patented (U.S. Patent 6,285,999). However, the patent is assigned to Stanford University and not to Google. Google has exclusive license rights on the patent from Stanford University. The university received 1.8 million shares of Google in exchange for use of the patent; the shares were sold in 2005 for $336 million." – Wikipedia.org

Reference: http://en.wikipedia.org/wiki/PageRank

Paid Inclusion

Paid inclusion is when an advertiser exchanges money for inclusion (not placement) on a search engine. It just gets you into the search engine's index but not a specific rank or placement within the search results.

Paid placement on the other hand…

Paid Placement

Paid placement is paying a search engine to have your listing show up with the search results pages. These listings are usually denoted as "sponsored" and appear in a special place or are somehow augmented (highlighted, etc).

Pay-For-Performance

Pay-for-performance is an advertising pricing model based on delivering quantifiable sales that can be directly attributed to the bottom line.

Pay-Per-Click (PPC)

Pay per click is a pay-for-performance pricing model where advertising is priced based on number of actual clicks that listing receives, rather than impressions or other criteria.

Pay-Per-Post (PPP)

Pay-per-post is when a content creator gets paid for each post they make on a website or blog. There are special websites which are designed to help content creators (bloggers) find advertisers willing to sponsor pay for content posted.

Phrase Match

Phrase Match is a form of keyword matching where an ad will be displayed if the user's search includes the exact phrase, even if their search contains additional words.

Portal

A portal is a website that functions as a point of access to information on the web. "Portal sites" such as travel websites, were a huge part of the DotCom Bubble of the late '90's and early 2000's.

Query

A query is another term for a search. When used as a verb, it is basically performing a keyword based search in a search engine.

When used as a noun, it refers to the actual keyword phrase the searcher types in to the search engine.

To keep it simple, you can just think of a query as a keyword or the act of entering a keyword into a search engine.

Reciprocal Linking

Reciprocal linking is the practice of trading links between websites. There is very little SEO value in reciprocal linking these days, although it was very worthwhile in the late 1990's until link farms made any reciprocal link seem irrelevant to search engines.

Redirect (or Auto Redirect)

A redirect, or more specifically an auto redirect, is achieved when the site visitor is automatically taken to another web page without actually clicking on a link.

Referrer

A referrer is simply a web page (or email message, etc.) that delivers a visitor to your site.

Relevance

Relevance can refer to several things in SEO. Most common are (1) How pertinent your keywords are to the web page or site theme. And (2) How applicable are the keywords on your page to those the user is actually searching for.

Repeat Visitor

A repeat visitor is one who accesses a website more than once over a specific period of time (typically measured in days or weeks).

Resubmitting

Resubmitting is the act of entering your URL to a search engine's submission feature or page, after you've already submitted it in the past, or after the search engine has already included your site in its index.

This is OK to a point, but overdoing it can get you in trouble. In fact, most search engines now have a stated clause that warns against resubmission.

Robot

See "Spider."

Robots.txt

Robots.txt is a text file placed in a websites root directory and linked to the html code. It allows for a person who is optimizing the site to control the actions of search engine spiders on the site or even deny them access.

ROI - Return on Investment

ROI or return on investment is the benefit gained in return for the cost of investing into advertising or a project, such as an SEO campaign. It can be measured easily as: "total revenues (generated from campaign or project) minus total costs."

This is more a business term but it's included here because anyone doing SEO or other online marketing, should focus on maximizing ROI – on every effort they undertake.

Run of Site (ROS)

Run of site is an Internet advertising program that delivers banner ads on any page of a website that rotates ads.

Scraper Sites

Scraper sites are designed to "scrape" search engine results pages or other sources of content to create content for a website.

Search Engine

A search engine is a website that offers visitors the ability to search the content of numerous web pages on the Internet.

Search engines periodically explore all of the pages of a website (that are easily accessible) and add the text on those pages into a large database that users can search.

Search Engine Marketing (SEM)

SEM, or search engine marketing, is basically any online advertising or publicity method that harnesses the power of search engines as the advertising medium. It includes, but is certainly not limited to, search engine optimization (SEO), pay-per-click (PPC), paid space placement, and contextual advertising.

Search Engine Optimization (SEO)

Search engine optimization is the use of strategies and tactics to increase the organic or natural search rankings of web pages in the search engines.

Search engine optimization involves multiple steps including technical website optimization (coding and tags), content optimization (copywriting), and link building.

Search Engine Results Page (SERP)

A search engine results page, or SERP, is a web page containing search results from Google, Bing, Yahoo, etc. SERPs are more commonly known as organic rankings or simply called search results.

Search Term

A search term is a keyword or phrase used to conduct a search engine query. Back in the day, these were often called queries.

Search Term Popularity

See "Keyword Popularity."

SEM

See "Search Engine Marketing."

SEO

SEO is an acronym for "search engine optimization" and/or "search engine optimizer". A *search engine optimizer* is someone who performs search engine optimization.

See "Search Engine Optimization" for more info.

SERP or SERPs

SERP is an acronym for "search engine results page(s)".

Session

See "User Session."

Sniffer Script

A sniffer script is a "lite" program (or script) that detects which web browser software a site visitor is using and then serves up the particular browser-specific cascading style sheet or page content to match.

Although originated by the web development community, these scripts have found their way into some SEO establishments.

Spam

Spam, as it relates to SEO, is the practice of using a variety of manipulation techniques that violate search engines user agreements. These include keyword stacking, keyword stuffing and spamdexing (see below).

Spamdexing

Spamdexing involves several methods but the most common is stuffing multiple keywords into unrelated blocks of text and cramming them together on a single web page to get higher search rankings, without adding relevant content.

Spamglish

Spamglish is keyword-rich gibberish in any language used as search engine fodder instead of thoughtfully written, interesting content. Spamglish often includes meaningless sentences and keyword repetition.

Spamming

Spamming is most commonly associated with the act of sending unsolicited commercial email. However, in the context of search engine optimization, spamming refers to using dishonest tactics to achieve high search engine rankings.

Spider

A spider is basically a robot that is sent out by search engines to catalog websites on the Internet. When a spider indexes a particular website it is known as being "spidered".

Spider Trap

A spider trap is an infinite loop that a spider may get caught in if it explores a website where the URLs of pages keep changing.

Splash Page

A splash page is a home page that typically lacks content and is meant to add a "Wow" factor for the visitor. Often times, splash pages are created in Flash. Splash pages are an annoyance to

Internet users as they are an extra hoop that the user has to jump through before they get to any meaningful content.

Splash pages are also damaging to search engine rankings because your home page (now a splash page) is typically considered by search engines as the most important page of your site.

Standards Compliant

Being standards compliant means that your site (1) Uses valid XHTML and CSS and (2) Separates the content layer from the presentation layer.

Because standards compliant sites are accessible and usable to both humans and spiders alike, they tend to rank better in search engines than non-compliant sites.

Static

Static refers to the nature of a non-dynamic web page or simply content that remains constant.

Static pages are typically created in some sort of content management system or WYSIWYG (what you see is what you get) editor like Dream-weaver or MS Expression web.

Of course, as much as most search engines like to see fresh content, having some static content can be a very good thing as it adds an element of stability to both the search engine and the visitor.

Stemming

Stemming is a process search engines use to deliver results based on a word's root spelling. An example would be similar search results returned for "clothes" as for the word "clothing."

Stop Character

A stop character is special character most often used in a gibberish machine-generated URL.

Typical stop characters include the ampersand (&), equals sign (=), and question mark (?).

Eliminating stop characters from all URLs on your website will go a long way in ensuring that your entire site gets indexed by Google.

Stop Word

A stop word is any one of a number of common terms, such as "the," "a," "an," "of," and "with". They are considered so common and meaningless that a search engine won't bother including them in their index, or database, of web page content.

Stop words on your web pages are treated as if they do not exist.

Tip: Consider how many stop words you have when you are calculating keyword weight. It can make a big difference in the final factor..

Streaming Media

Streaming media can be any form of audio-visual content that is played as it is being downloaded instead of the user having to wait until the entire file has downloaded to the computer. It became very popular in the mid 2000's as more people got high-speed Internet access.

Submitting

Submitting is the act of inputting a web page address to a search engine in the hopes that it will spider and then index it. Also, check out "resubmitting" for advice on how often to submit your web pages.

Supplemental Pages

Supplemental pages are basically pages held in reserve (yet indexed) by Google but do not typically appear in the SERPs at this time.

Target Audience

A target audience is the market in which advertisers wish to sell their product or service to.

Taxonomy

Taxonomy is a classification system of controlled vocabulary used to organize relevant subjects.

Technorati

Technorati is a search engine for blogs. If you want to find blogs around a certain topic or type you can search for a variety of blogs on Technorati.

Teoma

Teoma is one of the original Internet search engines. You don't hear much about it now, but it was a pioneer of the early web.

Term Frequency

The term frequency is the number of times a term occurs or repeats in a document. This is basically the same concept as keyword frequency but can relate to non-keyword terms as well.

Text Ad

A text ad is (typically) action oriented copy describing the product or service that is being advertised. The text ad appears alongside natural search results (but looks different from organic listings and is labeled as "sponsored") and links to a specified web page.

Text Link Ads

A text link ad is an advertisement in the form of a text link.

When you scroll over a text link ad the advertisement pops up from the link for you to click on. This kind of advertisement attracts visitors by putting the ad in a link (within the page content where it seems less "ad-like").

This is a great example of <u>anchor text</u> usage.

Theme

A theme can be most easily defined as the main keyword focus of a web <u>page</u>. The theme of a <u>website</u> relates to what the whole site is about or the kind of information the content is relaying.

Thesaurus

As you likely already know, a thesaurus is similar to a dictionary but contains a list of synonyms rather than definitions. You can look up words in a thesaurus to find similar words.

In SEO terms, many content "spinning" apps rely heavily on a digital thesaurus for its content pool. Nine times out of ten these apps produce junk that should be avoided at all costs.

Tip: An actual thesaurus however, can be a copywriter's best friend when trying to develop pages of similar yet unique content (as viewed by the search engines at least).

Title Tag

A title tag is a back-end bit of code that identifies the title of a web page to a browser. This is the text displayed in the blue bar at the very top of the browser window.

Note: As far as tags go, this is the one that search engines give the most priority to when ranking a page for a particular keyword.

Toolbar

A toolbar is a browser add-on that usually include a handful of functions and an option to search the Internet.

These can be quite handy for tracking an SEO campaign as you can quickly and conveniently get info like PR, age of website, how many pages have been indexed *, etc. (* available on special SEO toolbars.)

Trackback

A trackback is a notification that someone has linked to a document on your site. This enables authors to keep track of who is linking to their articles.

Tracking

Tracking is simply keeping "track" of your site. Tracking and reporting tools, such as Google Analytics, can help you can refine your ad placement options (and SEO campaign plan), as well as spending levels depending on the results.

Traffic

Traffic refers to the users (and number of) that visit a website.

Traffic Estimator

Googles AdWords Traffic Estimator is a tool that provides search volume, average cost-per-click, and position estimates for search advertising in Google's search results and content network.

TrustRank

TrustRank is a link analysis technique that separates useful web pages from spam.

Typepad

Typepad is a blogging service that hosts blogs and small business web sites.

Unethical SEO

Unethical SEO, or "black-hat" search engine optimization, is any technique of combination of techniques that are considered dishonest and can result in getting sites banned from the search engines.

Examples of this include keyword stuffing and hidden text or a combination of methods (For example: Making text on the page the same color as the background (thus hidden or invisible) and "stuffing" it, which typically consists of lists of keywords that are put there in the hope of tricking search engine spiders.

Unique Visitors

Unique visitors are a count of individual users who have accessed your web site. This is different from "hits," as hits count multiple visits from the same user.

URL

A URL is a web address. The acronym URL stands for uniform resource locator. They specify the exact location of a web page or any resource on the web, including documents, images, downloadable files, etc.

This is important in SEO as it relates to using keywords and other descriptive text within URL strings.

URL Rewrite

A URL rewrite is a technique used to help make website URL's more user and search engine friendly.

Usability

Usability is how user-friendly a website is. It generally relates to the navigation system, font size and color, background, etc.

User Agent

User agent is the name or designation of the spider that is currently visiting a page.

User Generated Content

User generated content is content created and published by a variety of end users online. Types of user generated content are wide and varied and may include videos, podcasts and posts on discussion groups, blogs, wikis, and social media sites.

User Session

A user session is an instance of an Internet user accessing your website for a length of time and then leaving. During a user session any number of pages may be accessed. A user session is "ended" once a specific period of inactivity has been exceeded.

Visibility

Visibility is how well-placed or how visible your website is in the search engines for relevant keyword searches. It generally refers to being well-ranked and easily found when people are searching for the products or services you offer.

Visit

A visit is any instance of a web user going to a web page.

Web browser

A web browser is software installed on the user's computer that allows them to view web pages. Popular ones include Microsoft Internet Explorer, Firefox, Safari, and Google's Chrome.

Web Crawler

A web crawler, also known as a robot or spider, is a program or automated script that "crawls" the web in a methodical manner following links and working with an indexing script to get all of that data into its search engine database or index.

Web Standards

web standards are standardized guidelines that help ensure that web sites are accessible on a wide variety of platforms and to a wide range of users including users with disabilities.

Web2.0

web2.0 refers to the new generation of web based services and communities characterized by participation, collaboration, and the sharing of information among users online. web2.0 includes things like wikis, blogs, and social networking sites which encourage user generated content and social interaction online.

Weblog

A weblog, or what is commonly referred to as a blog, is an online journal. Blog authors choose whether to blog openly or anonymously. weblog entries are usually made regularly and chronologically. The range of topics covered is endless.

Some weblogs focus on a particular subject like travel, fashion, or astrology while others are personal online diaries. weblogs are usually made up of posts, images, videos, comments and links. Popular blogging platforms include: Blogger, WordPress, Typepad, LiveJournal and Dreamhost.

Blogs are great for SEO purposes due to the ever-fresh content they deliver.

White Hat SEO

Just the opposite of "black hat", white hat SEO is considered ethical and approved by the search engines.

Wikipedia

Wikipedia is the largest and fastest growing encyclopedia online and is one of the world's ten most visited websites. Wikipedia is multilingual and currently available in 253 languages.

Operated by the Wikimedia Foundation, a non-profit organization, it is written collaboratively by volunteers around the world. It is a peer-reviewed publication, does not require contributor's legal names, and requires that contributions are supported by published and verifiable sources. Studies have shown that generally the site is as accurate as other encyclopedias.

WordPress

WordPress is an open-source web publishing and content management system. It was created primarily as blogging software and is ideal for managing content that is frequently updated.

WordTracker

WordTracker is a popular keyword research tool. It is designed to assist search marketing professionals and webmasters in identifying important keywords and phrases relevant to their website. It provides detailed information on the number of searches, predicted number of daily searches, competing pages, and keyword efficiency data. Information can be broken down by each search engine.

XML

XML is an acronym for extensible markup language. It is a scripting language interpreted by web browsers. XML is similar to, but more advanced than, HTML.

Yahoo!

Yahoo is the original friendly Internet search system. Until the early 2000's it served solely as a directory (a collection of Internet sites organized by subject, entered by site managers, then reviewed by human editors) rather than a true search engine, which send out a spider to crawl as many documents as possible and an indexer that reads each page or document and generates an index based on the information within each document.

Today, it has both a search engine and directory, with the search engine serving up the default SERPs (search engine results page). And, although it costs $299 to apply for listing in the directory, it is a wise investment for serious online businesses as it will add credibility to your site from any search engine's perspective.

Note: In late August 2010, Yahoo! began showing Bing's search results. Yahoo! is still Yahoo! but the results are Bing's – although actual placement on the page can vary (stay tuned for how this shakes out.)

Some History: Originally created in 1994, by Jerry Yang and David Filo, as a way to keep track of interesting links on the Internet, Yahoo was first called "Jerry and David's Guide to the World Wide Web."

They ran it from their personal computers at Stanford University while completing their PhD's in electrical engineering. Of course that name was a bit long so they came up with an acronym that had a double meaning - Yahoo! is short for "Yet Another Hierarchical Officious Oracle," which happens to be a word they liked the general definition of; yahoo: "rude, unsophisticated, or uncouth."

About the Author

In 1982 when Mike began his career in advertising, he was moved by these words from Zig Ziglar, a legendary salesman and sales trainer: "You will get all you want out of life if you help enough other people get what they want." Nearly thirty years later these words still ring true. For Mike it's a good day when he has helped someone else be successful - and he has a lot of good days.

With over a quarter century in direct marketing, success has come to Mike in several ways. He is the CEO and President of IMPAK Marketing, one of the southeast's largest direct marketing firms. Then in 2006, he built on this direct-marketing success by founding NetSearch Direct; a Search Engine Optimization and Marketing/Social Media company headquartered in Richmond Virginia.

From a small shop with a handful of specialists NetSearch Direct managed to double in size during the worst economy in 30 years. Thanks to some great successes, lucky breaks and plenty of hard work NSD now provides profit-producing results to over one hundred businesses throughout the United States.

Mike is the proud father of two young-adult daughters and firmly believes that whatever success has come his way has been primarily because of hard work, luck, more hard work and luck, and support from his wife of twenty-six years, Helene.

Free Search Engine Snapshot Report

NetSearchDirect.com   Fax to: 804.228.4479

Name: _____

Company: _____

Email: _____

Phone: _____ **Fax:** _____

Please complete this info and fax it to 804.228.4479 (or email answers to mikem@netsearchdirect.com). Once received we can have a **Free Search Engine Snapshot Report** back to you within a few days, then answer any questions you might have.

Web Address: _____

Keywords (What words or 'search terms' do you want to be found for?)

_____   _____   _____

_____   _____   _____

GEOs (What state and cities do you want to be found for?)

_____   _____   _____

That's it. Please fax to 804.228.4479. We'll be in touch soon. Thanks!